Azure Stack Hub Demystified

Building hybrid cloud, IaaS, and PaaS solutions

Richard Young

BIRMINGHAM—MUMBAI

Azure Stack Hub Demystified

Group Product Manager: Rahul Nair
Publishing Product Manager: Preet Ahuja
Senior Editor: Shazeen Iqbal
Content Development Editor: Romy Dias
Technical Editor: Shruthi Shetty
Copy Editor: Safis Editing
Project Coordinator: Shagun Saini
Proofreader: Safis Editing
Indexer: Manju Arasan
Production Designer: Vijay Kamble

First published: September 2021

Production reference: 1010921

Published by Packt Publishing Ltd.
Livery Place
35 Livery Street
Birmingham
B3 2PB, UK.

ISBN 978-1-80107-860-3

www.packt.com

To my wife, sons, and grandchildren, for their unwavering support

Contributors

About the author

Richard Young has been working in IT for over 35 years and is currently a principal consultant in the professional services division of Lenovo Global Technologies, covering EMEA. He works closely with customers to assist them in their journey to the cloud with a focus on hybrid cloud, especially focused on Microsoft Azure. His role covers both the strategy and deployment of hybrid cloud using Microsoft Azure Stack Hub and Azure Stack HCI. He holds both MCSE and MCSA certifications for Azure. He holds the MCPD certification for .NET development for the cloud, from back when he was a developer. He has been involved in multiple deployments of Azure Stack Hub throughout Europe over the last few years. He is a husband, father, and grandfather.

I would like to thank my wife, Jackie, and my family, for giving me the encouragement to undertake the writing of this book as a first-time author. I would also like to thank the Packt editing team for their professionalism and dedication during this whole process, which made it a less daunting undertaking.

About the reviewer

Jean-Benoit Paux has more than 15 years of experience in the IT industry, particularly in the areas of support engineering, Microsoft infrastructure consulting, and cloud architecture. He assists various clients in defining their Azure cloud strategy, building their cloud foundations, and implementing security controls. Over the last few years, he has also succeeded in applying his expertise by building hybrid architectures using Azure and Azure Stack portfolio products. As a Microsoft Certified Trainer, Jean-Benoit enjoys sharing his knowledge on different IT matters, including teaching Azure and Microsoft 365 courses.

I'd like to thank this book's author and Packt Publishing for the opportunity to review this book. It involved many good debates and energy to make this book. I would also like to thank the Microsoft Azure Stack support engineers and PM who I had the opportunity to work with, who taught me a lot about the internals, and sometimes worked days and nights with us to troubleshoot this wonderful product.

Table of Contents

3

Azure Stack Hub Deployment

Section 2: Identity and Security

4

Exploring Azure Stack Hub Identity

5

Securing Your Azure Stack Hub Instance

6
Considering DevOps in Azure Stack Hub

Section 3: Features

7
Working with Resource Manager Templates

8
Working with Offers, Plans, and Quotas

Section 4: Monitoring, Licensing, and Billing

12

Monitoring and Managing Azure Stack Hub

13

Licensing Models in Azure Stack Hub

14

Incorporating Billing Models

15

Troubleshooting and Support

Other Books You May Enjoy

Index

Preface

Azure Stack Hub is the on-premises offering from Microsoft that provides Azure Cloud services from the customer's own data center. It provides consistent processes between onsite operations and the cloud, allowing developers to test locally and deploy to the cloud in exactly the same manner.

Azure Stack Hub Demystified provides Azure Stack Hub administrators and operators with the skills needed to successfully deploy, configure, and maintain a hybrid cloud environment based on Azure Stack Hub and public Azure. It covers all of the features, functions, and services available within Azure Stack Hub and explains how they can best be used to create services for customers to consume.

Who this book is for

This book is aimed at Azure administrators and Azure Stack Hub operators who provide or are looking to provide cloud services to their end customers within their own data center through the use of Azure Stack Hub. This book will also be beneficial to those who are studying for the AZ-600 Configuring and Operating a Hybrid Cloud with Microsoft Azure Stack Hub exam. An understanding of Azure, plus some knowledge of virtualization, networking, and identity management, is recommended for those wishing to get the most out of this book.

What this book covers

Chapter 1, *What Is Azure Stack Hub?*, provides an introduction to Azure Stack Hub and how it fits into the Azure ecosystem, alongside a detailed understanding of typical usage scenarios for Azure Stack Hub.

Chapter 2, *Azure Stack Architecture*, details the building blocks and the concepts that underpin Azure Stack Hub infrastructure and architecture.

Chapter 3, Azure Stack Hub Deployment, details the deployment process and all the necessary prerequisites that need to be in place for a successful deployment, including critical decision points while planning a deployment of Azure Stack Hub.

Chapter 4, Exploring Azure Stack Hub Identity, provides an in-depth look at the identity management options for Azure Stack Hub, including Azure Active Directory and Active Directory Federation Services.

Chapter 5, Securing Your Azure Stack Hub Instance, covers Azure Stack Hub infrastructure security and compliance, including best practice approaches for securing an Azure Stack Hub instance and tackling breaches if they occur.

Chapter 6, Considering DevOps in Azure Stack Hub, details the concepts and practices for DevOps and how they relate to Azure Stack Hub, including details of the DevOps pipeline.

Chapter 7, Working with Resource Manager Templates, explains the authoring and deployment of Azure Resource Manager templates with Azure Stack Hub using Visual Studio Code, GitHub, and PowerShell.

Chapter 8, Working with Offers, Plans, and Quotas, explains how to create and offer services and resources from Azure Stack Hub to end customers via a subscription.

Chapter 9, Realizing Azure Marketplace, provides an overview of the Azure Marketplace and how it relates to Azure Stack Hub, including syndication and offline use.

Chapter 10, Interpreting Virtual Networking, details infrastructure as a service, including offering network services from Azure Stack Hub, with a focus on software-defined networking.

Chapter 11, Grasping Storage and Compute Fundamentals, provides an in-depth look at the storage and compute features available in Azure Stack Hub, including storage accounts, blobs, virtual machines, and more.

Chapter 12, Monitoring and Managing Azure Stack Hub, explains how to manage and monitor an Azure Stack Hub instance, including integration with common tools such as System Center Operations Manager.

Chapter 13, Licensing Models in Azure Stack Hub, provides a walk-through of the different licensing options for Azure Stack Hub for Windows and SQL running on the virtualization platform of Azure Stack Hub

Chapter 14, Incorporating Billing Models, details how consumption is tracked and billed across the different billing options in Azure Stack Hub, including the different metering tools and connection to the Azure public cloud subscription.

Chapter 15, Troubleshooting and Support, covers the integrated support model and troubleshooting tips for common problems seen in Azure Stack Hub.

To get the most out of this book

An understanding of Azure cloud services, including virtualization, networking, and identity management, would be beneficial to get the most out of this book. An appreciation of Active Directory is also recommended.

Software/hardware covered in the book	Operating system requirements
Azure Stack Hub	Windows (Microsoft Edge / Google Chrome)

If you are using the digital version of this book, we advise you to type the code yourself or access the code from the book's GitHub repository (a link is available in the next section). Doing so will help you avoid any potential errors related to the copying and pasting of code.

Code in Action

The Code in Action videos for this book can be viewed at `https://bit.ly/3yqknuF`.

Download the color images

We also provide a PDF file that has color images of the screenshots and diagrams used in this book. You can download it here: http://www.packtpub.com/sites/default/files/downloads/9781801078603_ColorImages.pdf.

Conventions used

There are a number of text conventions used throughout this book.

Code in text: Indicates code words in text, database table names, folder names, filenames, file extensions, pathnames, dummy URLs, user input, and Twitter handles. Here is an example: "For example, the IPv4 settings could be 10.128.0.28/30 or 10.128.32/30."

A block of code is set as follows:

```
# Create a PEP Session
winrm s winrm/config/client '@{TrustedHosts= "<IP_address_of_
ERCS>"}'
$PEPCreds = Get-Credential
$PEPSession = New-PSSession -ComputerName <IP_address_of_
ERCS_Machine> -Credential $PEPCreds -ConfigurationName
"PrivilegedEndpoint"
```

When we wish to draw your attention to a particular part of a code block, the relevant lines or items are set in bold:

```
params = @{
    ComputerName = $ErcsNodeName
    Credential = $CloudAdminCred
    ConfigurationName = "PrivilegedEndpoint"
}
```

Any command-line input or output is written as follows:

```
$cred = Get-Credential
Enter-PSSession -ComputerName <IP address of ERCS>
-ConfigurationName PrivilegedEnpoint -Credential $cred
Register-CustomDnsServer -CustomDomainName "<domain name>"
-CustomerDnsIPAddresses "ip1,ip2"
```

Bold: Indicates a new term, an important word, or words that you see on screen. For instance, words in menus or dialog boxes appear in **bold**. Here is an example: "Choosing **Do not connect to Azure** does not strictly mean that you cannot connect to the Azure Stack Hub instance to Azure for hybrid scenarios for tenant workloads."

> **Tips or important notes**
> Appear like this.

Get in touch

Feedback from our readers is always welcome.

General feedback: If you have questions about any aspect of this book, email us at customercare@packtpub.com and mention the book title in the subject of your message.

Errata: Although we have taken every care to ensure the accuracy of our content, mistakes do happen. If you have found a mistake in this book, we would be grateful if you would report this to us. Please visit www.packtpub.com/support/errata and fill in the form.

Piracy: If you come across any illegal copies of our works in any form on the internet, we would be grateful if you would provide us with the location address or website name. Please contact us at copyright@packt.com with a link to the material.

If you are interested in becoming an author: If there is a topic that you have expertise in and you are interested in either writing or contributing to a book, please visit authors.packtpub.com.

Share Your Thoughts

Once you've read *Azure Stack Hub Demystified*, we'd love to hear your thoughts! Scan the QR code below to go straight to the Amazon review page for this book and share your feedback.

https://packt.link/r/1801078602

Your review is important to us and the tech community and will help us make sure we're delivering excellent quality content.

Section 1: Architecture and Deployment

After reading this section of the book, you should have a complete understanding of the architecture of Azure Stack Hub and be able to describe the deployment process in detail.

The following chapters will be covered under this section:

- *Chapter 1, What is Azure Stack Hub?*
- *Chapter 2, Azure Stack Architecture*
- *Chapter 3, Azure Stack Hub Deployment*

1

What Is Azure Stack Hub?

This first chapter will introduce you to **Microsoft Azure Stack Hub** and how it is positioned within the **Microsoft Azure ecosystem**. You will gain a detailed understanding of the typical usage scenarios for Microsoft Azure Stack Hub and the Azure capabilities that are provided by the platform. We will cover the initial core fundamentals to prepare you for later chapters in this book. We will also cover the skills you will be tested on if you are looking to take the **Microsoft AZ-600: Configuring and Operating a Hybrid Cloud** with *Microsoft Azure Stack Hub exam*.

In this chapter, we're going to cover the following main topics:

- Introducing Azure Stack
- Understanding hybrid use cases
- Introducing Azure Arc
- Learning about Azure Stack integrated systems
- Exploring the AZ-600 exam requirements

Let's dive into the first topic.

Introducing Azure Stack

To begin this book, I thought the best place to start would be with a basic understanding of Microsoft Azure Stack Hub. The idea of this is to look at a question I am asked by customers all the time. *What is Microsoft Azure Stack Hub?* In simple terms, then, Microsoft Azure Stack Hub is an extension of Microsoft Azure, but this is only part of the answer. Microsoft Azure Stack Hub is a **hybrid cloud platform** that allows you to use Azure services from your company or a service provider data center. When people think of Microsoft Azure, they think of the *public cloud* offered by Microsoft, but it is, in fact, a complete ecosystem that incorporates not just the public cloud but also the on-premises versions called **Microsoft Azure Stack**. This includes Microsoft Azure Stack **HCI**, which stands for **Hyper-Converged Infrastructure**. HCI will be explained in detail in *Chapter 2, Azure Stack Architecture* but for now, it is enough to say that with HCI, both compute and storage are supplied from the same server. This is different from a traditional infrastructure, where storage and compute are separate. Microsoft Azure Stack **Edge** along with Microsoft Azure Stack Hub and Microsoft Azure Stack HCI conform to this pattern. This book is only focused on Microsoft Azure Stack Hub, but it is worth understanding the complete ecosystem as this will help highlight the differences between the different versions of solutions under the Microsoft Azure Stack banner. This becomes important especially when it comes to running solutions in a hybrid cloud scenario, which we will cover later in this chapter. The advantage of Microsoft Azure Stack is that it provides a consistent environment that those who already use Microsoft Azure will be more than familiar with. In fact, the promise of Microsoft Azure Stack Hub when you talk it through can be thought about in terms of the following concepts:

- Consistent application development
- Azure services available on-premises
- Integrated delivery experience

For a developer who builds cloud applications for Microsoft Azure, they can take all the skills and tools they already use onto this platform. The deployment process that's used for Microsoft Azure is the same one that's used for Microsoft Azure Stack Hub. Development tools such as *Visual Studio* can also be used within this environment. Microsoft markets the fact that applications that run in Microsoft Azure can be run on Microsoft Azure Stack Hub with no changes other than deployment location, which is not strictly the case as some changes are nearly always required.

Microsoft Azure capabilities are also available within Microsoft Azure Stack Hub, which, again, breeds familiarity both from a developer standpoint but also from an operator and administrator standpoint. The following Microsoft Azure capabilities can be found in Microsoft Azure Stack Hub:

- **Virtual machines**: Rapid deployment with scaling on demand.

- **Containers**: Linux and Windows Servers containers, Azure Kubernetes Services.

- **Networking**: Virtual Network, Load Balancer, VPN Gateway, network security groups, public IPs, route tables.

- **Storage**: Blobs, tables, and queues.

- **Key Vault**: Securely protect application keys and secrets.

- **Azure App Service**: Web and API applications, Azure Functions, serverless computing.

- **Azure Marketplace**: Ready to go applications from the Azure Marketplace.

- **Event Hubs**: Scalable event processing for ingesting and processing large amounts of event data.

- **Azure IoT Hub**: Centralized message hub for communications between IoT applications and devices.

We will be covering each of these capabilities and services in detail later in this book, along with their limitations, as they are integral to creating offers and services from Microsoft Azure Stack Hub.

Supporting the Azure Stack Hub infrastructure

In addition to Microsoft Azure's capabilities and the support offered by Microsoft, Microsoft Azure Stack Hub is also supported by a myriad of both *hardware* and *software* vendors. I myself work for *Lenovo*, who provide certified hardware solutions that can be used to run Microsoft Azure Stack Hub on-premises, and I also work closely with Microsoft to ensure they adhere to the best practices when it comes to deploying Microsoft Azure Stack Hub. Lenovo are by no means the only hardware vendor to offer certified hardware for Microsoft Azure Stack Hub, and it is also supported on offerings from *Dell*, *HPE*, and *Cisco*, among others.

As well as the various hardware solutions that are available in the market, Microsoft Azure Stack Hub is also supported by software vendors extensively. Some industry standard solutions that are available to run in Microsoft Azure are also supported in Microsoft Azure Stack Hub through the Azure Marketplace. This allows customers to run the same software applications, such as *Red Hat*, *F5*, *Docker*, *Kubernetes*, *Chef*, and so on, in the same way in both their on-premises environment and the public cloud via Microsoft Azure.

Given the support of Microsoft and their hardware partners, this allows Microsoft Azure Stack Hub to offer a fully integrated delivery experience. Microsoft Azure Stack Hub is fast to deploy, allowing customers to get up and running quickly. The billing model within Microsoft Azure Stack Hub can be extended from Microsoft Azure to allow you to pay for use within the same Microsoft Azure subscription bill.

The key takeaway for Microsoft Azure Stack Hub from this quick overview is that this is an on-premises version of Microsoft Azure that is fully owned and operated by the customer within their own data center. Customers completely control the access, applications, and data that's stored in their Microsoft Azure Stack Hub. They are also responsible for ensuring that any applications or data being provided by Microsoft Azure Stack Hub are available at all times to their customers, regardless of whether they're internal or external. Therefore, I always describe Microsoft Azure Stack Hub as your own private Microsoft Azure region and you as the operator performing the role of Microsoft.

The real power of Microsoft Azure Stack Hub is when it is combined with the public Microsoft Azure Cloud in a truly hybrid manner. Throughout this chapter, I will introduce you some common hybrid use cases that I come across when I am working with customers during their cloud journeys.

Microsoft Azure Stack Hub is really the only consistent hybrid cloud where the tools and processes are consistent. Not only are the tools and processes consistent but so is the underlying infrastructure. As an example, let's take a look at some of these and why they work so well when it comes to running a hybrid cloud environment.

The following diagram tries to illustrate that Azure and Azure Stack Hub are consistent in the way they present their tools and processes:

Figure 1.1 – Consistency of tools and processes

We now have a clearer picture of what Microsoft Azure Stack Hub is and how it is closely related to Azure. To prepare you for the next chapter, we will now dive into how Microsoft Azure Stack Hub is used in disconnected scenarios for private cloud.

Understanding private cloud

Azure Stack Hub can be deployed in two different scenarios, depending on whether connectivity to Azure is required or not. One of the attractions of Azure Stack Hub is that it can be run completely standalone, with no connectivity to the internet. This is particularly useful for organizations that want the capabilities that are offered by the cloud but are unable to make use of public cloud offerings. This may be due to regulatory restrictions on data storage, latency issues with connectivity to public Azure, secure environments with no internet connectivity, environments with limited or unreliable network connectivity, and more.

The other use case is where you have a disconnected instance of Azure Stack Hub running in your data center. This is for organizations that are looking to modernize their applications on-premises and have legacy applications that cannot be moved into the public cloud.

Edge and disconnected solutions

Microsoft Azure Stack Hub can be used for applications where there may be connectivity issues in edge locations with limited network bandwidth. This allows logic and data processing to be performed closer to the users. This also applies to locations where *real-time latency* may be a consideration. An example of this I have seen was with a customer I have worked with who was capturing telemetry from trains. This can also equally be applied to locations such as oil rigs, cruise ships, or secure government sites.

Azure Stack Hub is not just beneficial as a private cloud but is also a key part of a hybrid cloud. We will look at some of these use cases next.

Understanding hybrid use cases

Microsoft provides a unified development and *DevOps* environment between their **Microsoft Azure cloud** offering and Microsoft Azure Stack Hub. Using tools such as *Visual Studio Team Foundation*, GitHub, and Azure DevOps, developers can work with the same processes, regardless of where their code is ultimately published to.

Microsoft Azure and Microsoft Azure Stack Hub share a common identity model. The on-premises Microsoft Azure Stack Hub utilizes Azure **Active Directory** but can also use *Active Directory Federation Services*.

Microsoft Azure and Microsoft Azure Stack Hub also share an integrated management and security control platform as both use the same Azure portal. This allows operators and administrators to ensure that access controls are consistent through the use of role-based access control.

They both share a common and consistent data platform, which is based on *a storage account that is used to provision Blobs/Tables/Queues that are available in both Azure and Azure Stack Hub.*

The following diagram shows the common set of functionalities that are shared between both the Microsoft Azure Stack on-premises and the public Azure cloud:

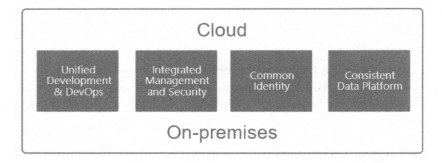

Figure 1.2 – Common functionality

Now that we've looked at the functionality that is shared between on-premises Azure Stack Hub and the Azure public cloud, we can delve into some common use cases for this hybrid adoption, starting with development.

Development

When I work with customers who build applications or services that are designed to run in the cloud, be that *Microsoft Azure*, *Google Cloud*, or *Amazon Web Services*, they tend to find that their development teams can rack up large costs by using the same cloud environment for their development environment, as well as production. The biggest attraction of using the public cloud for development is the fact that it is so easy to spin up an environment. Unfortunately, from a development standpoint, this is also where most of the costs start to come into play, as environments are spun up for a particular project or development team but are not always torn back down when they are finished with. By moving these development environments into an on-premises environment in Microsoft Azure Stack Hub, the customers can begin to make real savings as they have complete control over these environments. They can ensure that machines are removed once projects are completed.

The beauty of this for the developers is that they see no change in the tools or the process for building or deploying their code other than the endpoint. A developer can create their code in Visual Studio and deploy it directly from the **Integrated Development Environment (IDE)** to Microsoft Azure Stack Hub to test and **Quality Assure (QA)** it. Once tested and they are comfortable it works as designed, with no changes needing to be made, they can deploy the same release to Microsoft Azure public cloud.

For some organizations, the reverse of this is true and development is, in fact, done in the public cloud rather than on-premises. This is due to the flexibility offered by the public cloud and the speed with which environments can be spun up and down.

Testing

In a similar vein to the development environment, the same logic can also be applied to **test** and **QA environments**. Whether this is to test new services that are going to be deployed to the Microsoft Azure public cloud or changes to existing services that have already been deployed to the Microsoft Azure public cloud, then being able to test these in an environment that behaves in the same manner, but with no additional cost, is a great reason for running Microsoft Azure Stack Hub. Again, as with the development hybrid use case, these environments can be torn down once the release has passed testing and been released into production.

Regulatory

There are times where regulatory restrictions prevent data from being stored or manipulated in the public cloud. Theis can be dictated by government, industry, or regions. This may be because data cannot be stored in the public cloud or because data must be stored within the same country as the organization, and Azure is not available in that country. This is particularly true for multi-national companies who may have different regulations to contend with from different countries and governments, but they want to provide a consistent experience to all their employees. The idea of being able to develop and deploy global applications in Microsoft Azure for most locations, while still using the same deployment in local on-premises Azure Stack Hub where local restrictions dictate, is key. Application examples include *global audits, financial reporting, foreign exchange trading, inline gaming, health data*, and *expense reporting*.

Cloud application model

For customers running legacy applications, Microsoft Azure Stack Hub gives them the opportunity to apply modern architectures to their on-premises applications, which are not yet ready for the cloud. This brings into focus things such as containers and microservices, which can be tested on-premises in Microsoft Azure Stack Hub, safe in the knowledge that once they work in Microsoft Azure Stack Hub, they can then be deployed to Microsoft Azure with no code changes. Again, this is providing a consistent programming model, skills, and processes. You can use consistent processes across Azure in the cloud and Azure Stack Hub on-premises to speed up app modernization for core mission-critical applications. Azure Stack Hub is not simply just a virtualization platform such as Hyper-V or VMware; it is a fully fledged modern cloud platform.

Why is it compelling?

Organizations can now modernize their applications across hybrid cloud environments, balancing the right amount of flexibility and control. Developers can build applications using a consistent set of *Azure services* and *DevOps practices*, then collaborate with operations to deploy to the location that best meets their business, technical, and regulatory requirements. Developers can speed up new cloud application development by using pre-built solutions from the *Azure Marketplace*, including open source tools and technologies.

Note that this is all about applications. That is where the real value of a new hybrid cloud platform is. This will allow applications that are not yet ready to be run in a cloud environment to start moving in this direction. Cloud computing is likely to become the dominant design style for new applications and for updating many applications over the next 10+ years.

We have now covered the general hybrid use cases that can be undertaken on the Microsoft Azure Stack Hub platform. From here, we will take a look at one other use case that is not directly related to hybrid or private cloud scenarios, and that is Azure Arc.

Introducing Azure Arc

For true versatility, we need to look further than just the standard hybrid use cases we discussed in the previous section. This is where Microsoft have introduced *Azure Arc*:

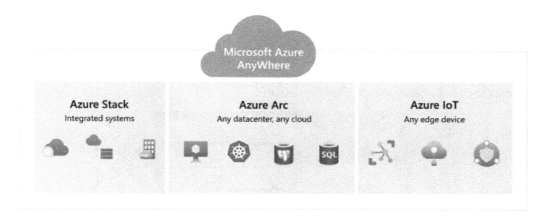

Figure 1.3 – Azure Arc

When taken with the Microsoft Azure ecosystem, Microsoft Azure Stack Hub starts to bring real mobility to the hybrid cloud. With the adoption of **Microsoft Azure Arc**, it is possible to run the same applications virtually anywhere, be that *on-premises*, at the *edge*, or on any *public cloud*. Azure Arc is a software solution that enables you to manage all your resources, including your on-premises resources, multi-cloud resources, virtual servers, and Kubernetes clusters from a single pane of glass as if they were all running within Azure.

Now, we will look at the integrated systems that are offered by the OEM vendors.

Learning about Azure Stack integrated systems

In this section, I will cover the Microsoft Azure Stack Hub **integrated systems**, which are only available from the hardware vendors who partner with Microsoft to certify their solutions can run Microsoft Azure Stack Hub. This includes the likes of Lenovo, Dell, HPE, and Cisco, among others. Azure Stack Hub cannot be built using normal servers from the vendors, and it is not possible to build an integrated system that's not supplied by one of the OEM vendors.

An Azure Stack Hub integrated system provides the *software*, *hardware*, *support*, and *services* needed in one fully supported platform.

To start, let's look at the standard infrastructure that is consistent across all the hardware vendors.

Why Hyper-Converged Infrastructure (HCI)?

In this section, we'll look at why infrastructure has evolved into the HCI in the modern data center, as it has with Microsoft Azure Stack Hub. To do this, we will start with a little bit of history of the evolution of the data center infrastructure. The following diagram represents the traditional three tier infrastructure:

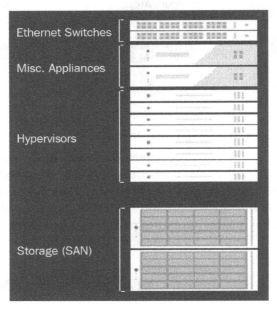

Figure 1.4 – Traditional infrastructure

The **traditional data center infrastructure** relied on specialized, discrete hardware components from compute, network, and storage or bare metal. These components were typically configured into silos of infrastructure to support specific workloads or applications. With traditional infrastructure, customers face challenges with integrating disparate infrastructure components, complex technical configuration, interoperability constraints, understanding the implications of the technology's architecture, and specialized administrative skills for compute, network, and storage technologies. IT teams must then coordinate across all these disciplines and operational domains to scale capacity, collectively provision resources and connectivity for applications, and manage updates and upgrades across this infrastructure.

The following diagram represents the change from the traditional infrastructure to the newer hyper-converged infrastructure that underpins Microsoft Azure Stack Hub:

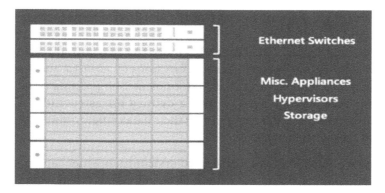

Figure 1.5 – HCI

Software-defined compute introduced consolidation technologies such as *server virtualization* and *containers* to reduce server sprawl where data centers suffered from overpopulation and complexity, by poorly utilized server hardware dedicated to application silos.

This compute consolidation helped optimize *server utilization* but led to additional stress and complexity in networking and storage administration. While this was an evolutionary step in managing compute resources, the balance of operational complexity materially shifted from compute to *networking* and *storage* domains, which remained highly specialized and independently managed.

Converged infrastructure brought more standardization to how software-defined compute was integrated with networking and storage technologies. While these technology domains continued to be operated separately, they could be delivered together as a standardized, integrated infrastructure that eliminated the traditional burdens of managing component interoperability, best practice architecture, and baseline configuration. Standardizing the infrastructure's design and integration provided incremental improvements to cross-functional operations and life cycle management.

HCI combines two or more *software-defined* components that are tightly integrated to be operated on as one common platform. The most popular form of HCI is to combine software-defined compute with **software-defined storage** (SDS), data management, and storage services implemented in software rather than dedicated hardware, which further reduces the operational overhead involved in managing and updating those technologies individually. This further simplifies infrastructure deployment due to the consolidation of multiple technologies in single appliances that can be clustered together. The addition of software-defined networking adds even more to this simplification by allowing all the components from a traditional architecture be managed from a single pane of glass. This truly brings Azure Stack Hub into the heart of data centers.

On-premise privates cloud delivers a service-oriented delivery, consumption, and operating model across a fully integrated, end-to-end automated infrastructure platform within a customer's data center. Cloud capabilities also typically include self-service controls, built-in facilities to offload application functions or services, and standardizing offerings in the form of a marketplace or catalog. Implementing private cloud capabilities is increasingly simplified with the use of **software-defined infrastructure** (SDI), and customers can leverage any degree of SDI in their data center to suit the level of operational agility they wish to achieve.

With Microsoft Azure Stack Hub as an integrated system, all updates can be applied across hardware, and both server and storage virtualization software at the same time. Microsoft Azure Stack Hub is easy to grow by simply adding extra nodes to the cluster, which expands both storage and compute capacity together. This removes the need to manage a separate storage system and SAN. A HCI such as Microsoft Azure Stack Hub embeds SDS and software-defined compute into an integrated single management experience.

An example of an OEM vendor integrated system is shown here:

Figure 1.6 – Lenovo ThinkAgile SXM for Azure Stack Hub

Azure Stack Hub is part of a family of products under the Azure Stack banner, as shown in the following diagram:

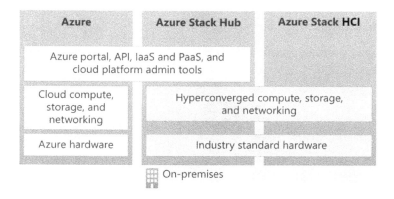

Figure 1.7 – Azure family of products

In addition to Azure Stack Hub, Microsoft also offers Azure Stack HCI, which is another member of the Azure Stack family. Azure Stack HCI is built upon Windows 2019 failover clustering, Hyper-V, and Storage Spaces Direct. Unlike Azure Stack Hub, the goal here is to provide simple virtual machine and container hosting while leveraging a public Azure cloud service for cloud backup or remote management. You will not have your local Azure region on-premises with the full breadth of services and user experience. Azure Stack HCI was originally based on Windows 2019 but is now available as an Azure service with subscription-based billing. It is based on the same core operating system components as Windows 2019 but is a new product line entirely, specifically focused on virtualization. Typical use cases for the Azure Stack HCI version are as follows:

- Remote or branch office
- Data center consolidation
- Virtual desktop infrastructure
- Lower-cost storage
- High availability and disaster recovery in the cloud

We will not be covering Azure Stack HCI in any more detail in this book as it is a different platform to Azure Stack Hub.

Now that we have an understanding of the history of hyper-converged infrastructure and the integrated systems, let's look at appliances.

Appliances

Appliances, like integrated systems, deliver Microsoft Azure consistent innovation with tightly controlled and thoroughly tested hardware/firmware/software combinations for the best reliability and availability.

The following diagram shows a standard Microsoft Azure Stack cluster running on certified appliances:

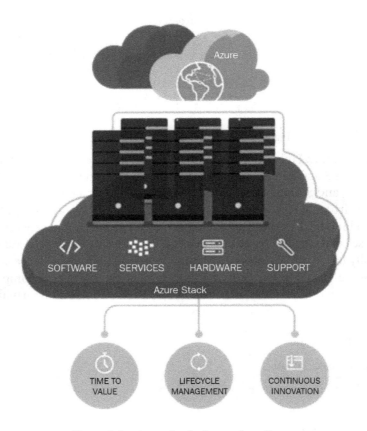

Figure 1.8 – Azure Stack clustered appliances

The Microsoft Azure Stack Hub certified appliances include everything needed to run Microsoft Azure Stack Hub, including *servers*, *BMC switches*, and *TOR switches*. All the hardware vendors offer full solutions as an appliance, which includes everything you would need to be able to run Microsoft Azure Stack Hub, once integrated into your data center. The whole solution must be purchased based on set configurations from the OEM vendors and must be purchased as a complete unit from only one OEM vendor.

Support

The integrated systems also deliver a consistent **support** experience, no matter who the customer contacts for support. There are coordinated escalation and resolution processes in place, with the same ticket being passed between the hardware *OEM* vendor and Microsoft. The appliance is supported by the hardware OEM vendor, while the associated cloud services are supported by Microsoft, who have back-to-back agreements with each of the hardware vendors who offer Microsoft Azure Stack Hub *certified solutions*. All updates, policies, and tests are coordinated between Microsoft and the OEM vendor.

Minimum hardware requirements

Each Microsoft Azure Stack Hub appliance needs to adhere to these **minimum hardware requirements** to be certified by Microsoft. Each vendor ensures that their firmware and software stacks are compatible with these requirements.

For compute, you need the following:

- **CPU**: 20 cores minimum (2 sockets at 10 cores each)
- **Memory**: 256 GB
- **NIC**: 2-port 10 GbE or better
- **Boot device**: 400 GB or larger

For storage, you need the following:

- **Cache**: 2+ flash drives (NVMe, SATA SDD, SAS SDD)
- **Capacity**: 4+ capacity devices (HDD or SDD)

The allocated ratio of cache to capacity is generally set to 10%.

The **top of rack** (**TOR**) switches consist of two switches per scale unit, configured for resiliency with 10 GbE or better for server connectivity. The switches must be capable of supporting *BGP*, *DCB*, *PFC*, *ETS*, and *multi-chassis link aggregation*. A scale unit is the minimum configuration of four servers or nodes that are clustered together to form the base scale unit.

The BMC or management switch should be a 1 GbE switch capable of L3 routing and simultaneous connectivity to the TOR switches.

These switch devices are then clustered together with between 4-16 nodes to form the full Microsoft Azure Stack Hub solution. These will be dependent on the workload it will be running for the customers.

The following diagram shows an Azure Stack Hub integrated system with the minimum configuration of four nodes:

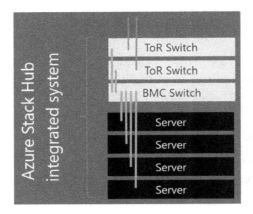

Figure 1.9 – Azure Stack Hub integrated system scale unit

Before we move on from this section, there are another couple of items to cover, starting with how to procure an Azure Stack Hub solution.

If you are an organization that is looking at implementing Azure Stack Hub, then the hardware must be purchased from an OEM vendor as an integrated system. You can choose the vendor you are most comfortable with, such as Lenovo, HP, Dell, Cisco, and so on, and they will have different configurations for you to choose from, depending on the workload you are planning to run. I know from my experience working for a vendor that a lot of the procurement of Azure Stack Hub is done through the RFP process. This allows the organization to define their requirements from a technology-agnostic standpoint and allows the vendor to define the correct configuration based on these requirements.

If you are unsure whether Azure Stack Hub is right for you, then Microsoft have a development version you can use for free to evaluate it, which is the Azure Stack Hub Development Kit. This development kit works against a single server, and any application that is built on here will work when it's deployed to a full Azure Stack Hub integrated system. The free Azure Stack Hub Development Kit is available to be downloaded from the Microsoft website. There are minimum hardware requirements for the Azure Stack Hub Development Kit, all of which are detailed here:

- 1 operating system disk with a minimum of 200 GB available
- 4 data disks each providing at least 240 GB capacity
- Dual-socket 16-physical-core processor
- 192 GB RAM

- Hyper-V enabled
- Windows Server 2019

This is a great option to try before you buy. Alternatively, most of the OEM vendors will also have either a demo kit or rental kit, which can be used for a proof of concept.

We now have a grounding in the Microsoft Azure Stack Hub infrastructure and how this is supported by OEM hardware vendors who build certified integrated systems. We understand the history of hyper-converged infrastructure, along with the benefits this brings. This gives us a good start in the fundamentals of the architecture, which we will build on throughout the rest of this book. I also want to take a moment to look at the AZ-600 exam, which we will run through in the next section.

Exploring the AZ-600 exam requirements

Along with our introduction to Microsoft Azure Stack Hub, I want to also run through the requirements for the **AZ-600 exam** from *Microsoft*. The **Microsoft Exam AZ-600: Configuring and Operating a Hybrid Cloud with Microsoft Azure Stack Hub**, is aimed at Microsoft Azure *administrators* or Microsoft Azure Stack Hub *operators* who are looking to provide cloud services to their end customers from their own data center. If you wish to pass the AZ-600 exam, it is worth noting the skills that are to be measured. The remainder of this book will work as an aid in preparation for this exam and will cover all the relevant skills that are to be measured.

The following skill measurements have been taken from the Microsoft exam website and are intended to illustrate how the skill is assessed. This is by no means an exhaustive list and will be subject to change by Microsoft over time.

Provide services (30 - 35%)

The first area to look at for the exam is the provision of services, which includes **Azure Marketplace** and its service offerings. This will account for *30 – 35%* of the exam:

- Manage Azure Stack Hub Marketplace:

  Populate Azure Stack Hub Marketplace in a disconnected environment

  Create a custom Azure Stack Hub Marketplace item

  Manage the life cycle for Azure Stack Hub Marketplace items

- Offer an App Services resource provider:

  Plan an App Services resource provider deployment

Deploy an App Service resource provider

Update an App Services resource provider

Scale roles based on capacity requirements

Rotate App Services secrets and certificates

Manage worker tiers

Back up App Services

- Offer an Event Hub resource provider:

Plan an Event Hub resource provider deployment

Deploy an Event Hub resource provider

Update an Event Hub resource provider

Rotate Event Hub secrets and certificates

- Offer services:

Create and manage quotas

Create and manage plans

Create and manage offers

Create and manage usage subscriptions

Change user subscription owner

- Manage usage and billing:

Set up usage data reporting

View and retrieve usage data by using the Usage API

Manage usage and billing in multi-tenant and CSP scenarios

Implement data center integration (15 – 20%)

For the exam, you also need to have an appreciation of the deployment process, especially when it comes to networking and certificates. This part of the exam is going to account for *15 – 20%* of the questions:

- Prepare for Azure Stack Hub deployment:

Recommend a name resolution strategy

Recommend a public and internal IP strategy

Recommend a data center firewall integration strategy

Recommend an identity provider

Validate identity provider integration

Configure the time server (NTP)

- Manage infrastructure certificates for Azure Stack Hub:

Recommend a certificates strategy

Validate the certificates

Run a secret rotation PowerShell cmdlet for external certificates

- Manage Azure Stack Hub registration:

Recommend a registration model

Register in a connected environment

Register in a disconnected environment

Re-register

Manage identity and access (10 – 15%)

As part of the AZ-600 exam, you will also need understand how to manage and configure access, which includes service principals. This will equate to *10 – 15%* of the questions you are likely to see when you take the exam:

- Manage multi-tenancy:

Configure the Azure Stack Hub home directory

Register the guest tenant directory with Azure Stack Hub

Disable multi-tenancy

Update the guest tenant directory

- Manage access:

Identify an appropriate method for access (service principal, users, and groups)

Provision a service principal for Azure Stack Hub

Recommend a permission model

Configure access in Azure Stack Hub

Create a custom role

Manage infrastructure (30 – 35%)

The final portion of the exam will focus on managing the **Azure Stack Hub infrastructure**, including capacity planning and monitoring health. It is likely to include questions around the update process and privileged endpoints. This portion of the exam will account for *30 – 35%* of the questions you will see in the exam:

- Manage system health:

    Recommend a monitoring strategy

    Monitor system health by using the REST API

    Include resource providers such as Event Hubs

    Monitor system health by using the Syslog server

    Manage field replacement or repair

    Configure automatic diagnostic log collection

    Collect diagnostic logs on demand by using PowerShell

    Configure Syslog forwarding for Azure Stack Hub infrastructure

- Plan and configure **Business Continuity and Disaster Recovery** (BCDR):

    Recommend a **BCDR** strategy

    Recommend a strategy for infrastructure backups

    Configure a storage target for infrastructure backups

    Configure certificates for infrastructure backups

    Configure a frequency and retention policy for infrastructure backups

- Manage capacity:

    Plan for system capacity

    Manage partitioned GPUs

    Add nodes

    Manage storage capacity

    Add IP pools

- Update infrastructure:

    Update Azure Stack Hub

    Download and import update packages manually

    Update Azure AD home directory

- Manage Azure Stack Hub by using Privileged Endpoints:

  Connect to a privileged endpoint

  Configure the Cloud Admin user role

  Unlock a support session

  Close the session on the privileged endpoint

  Stop and start Azure Stack Hub

  Perform system diagnostics by using Test-AzureStack

Summary

This first chapter has given us a brief introduction to Microsoft Azure Stack Hub. It has allowed us to understand that Microsoft Azure Stack Hub is an extension of Microsoft Azure that is run on-premises within a customers' data center. We have learned that it is considered an HCI platform that is supported by both hardware and software vendors. We now know it is a consistent hybrid cloud platform that offers Azure services that are integrated with both infrastructure as a service and platform as a service. We have also learned about the capabilities that can be exposed by the platform, which means we should be able to explain the hybrid use case scenarios for which Microsoft Azure Stack Hub can be utilized.

We should also be able to describe the minimum hardware requirements of the integrated systems provided by the hardware vendors. Finally, we looked at the **AZ-600: Configuring and Operating a Hybrid Cloud Platform with Microsoft Azure Stack Hub** exam and now understand what skills are measured as part of this exam.

In the remainder of this book, we will build on this foundation and cover each of the capabilities of Microsoft Azure Stack Hub in greater detail.

In the next chapter, you will dive into the underlying architecture that underpins Microsoft Azure Stack Hub and the building blocks of the platform.

2
Azure Stack Architecture

This second chapter follows on from the brief overview we provided in the previous chapter and will introduce you to the underlying architecture that constitutes the **Microsoft Azure Stack Hub** solution. This chapter will provide details around the building blocks and the concepts that underpin Microsoft Azure Stack Hub. By the end of this chapter, you should be able to describe the components that make up the underlying architecture of Microsoft Azure Stack Hub.

This chapter is a starting point and is purely an introduction to the building blocks of the Microsoft Azure Stack Hub, which we will continue to build on through the remaining chapters of this book. Understanding the basic building blocks will be important for those of you looking to take the **AZ-600 exam**.

In this chapter, we are going to be covering the following main topics:

- Providing an in-depth architecture overview
- Learning about capacity, scalability, and resiliency
- Understanding compute
- Getting to know the networking component
- Introducing storage

Let's begin with a more *in-depth overview of the architecture* for Microsoft Azure Stack Hub.

Providing an in-depth architecture overview

As we touched on in the previous chapter, the way you should purchase the infrastructure to run Microsoft Azure Stack Hub from an *OEM* hardware vendor is as an integrated system. These integrated systems include all of the software, hardware, support, and services necessary to run Microsoft Azure Stack Hub on-premises within the customer's own data center. But the infrastructure that is used to run Microsoft Azure Stack Hub is only part of the story when it comes to the **architecture**.

It is an important part of the story, and the process of implementing this architecture goes through a complete life cycle between the customer and their chosen hardware OEM vendor, as shown in the following diagram:

Figure 2.1 – The architecture design process

The OEM hardware vendors, such as *Lenovo, HPE, Dell*, and so on, work with their customers to explain the architecture, hardware, and topology of Microsoft Azure Stack Hub, which fits the requirements of the workload the customers are planning to run on their platform. This information is then fed into a deployment and configuration worksheet, which is, in turn, used to provision the Microsoft Azure Stack Hub within the customer's data center. This deployment is then validated with **PowerShell** scripts. We will be covering the deployment process in detail in a later chapter of this book. For now, all you need to know is that the architecture of Microsoft Azure Stack Hub is carefully designed and configured between Microsoft and their OEM hardware partners.

With this understanding of the architecture design process, we can move on to the **Cloud Operating Model**.

Cloud Operating Model

The starting point for the architecture of Microsoft Azure Stack Hub is the cloud operating model, which is identical to what is utilized by Microsoft in the public Azure cloud. Microsoft Azure customers use services and applications from the Azure public cloud, which, in turn, is operated by Microsoft engineers. The same is true of Microsoft Azure Stack Hub, where the organization's customers use services and applications from Microsoft Azure Stack Hub, which is operated by the IT engineers of the organization. Typical roles from IT who work with Microsoft Azure Stack Hub are **Cloud Architect** and **Cloud Operator**. The *Cloud Architect* will be involved in designing and developing the services to be supported on Microsoft Azure Stack Hub. The *Cloud Operator*, on the other hand, is responsible for ensuring that these services and applications are available from within Microsoft Azure Stack Hub. The Cloud Operator will, in this instance, not only be responsible for ensuring the applications and services are available, but also responsible for ensuring that the underlying infrastructure of Microsoft Azure Stack Hub is performant. The Cloud Operator effectively performs the same role on Microsoft Azure Stack Hub that Microsoft performs for the public Azure cloud.

Now that we have covered the cloud operating model and the different roles to be performed by the IT department, we can start to look at the **infrastructure software** underpinning Microsoft Azure Stack Hub.

Infrastructure software

We know that Microsoft Azure Stack Hub is considered a hyper-converged infrastructure, which means that its compute, storage, and networking is all contained within a single appliance. This is also referred to as **software-defined infrastructure** or **SDI**. So, the question for Microsoft Azure Stack Hub is, *what runs where?*

In an appliance, the physical servers running Microsoft Windows Server 2019 with Hyper-V are used to provide the compute, storage, and networking elements of Microsoft Azure Stack Hub. All the infrastructure roles within Microsoft Azure Stack Hub are hosted as **VMs**. This offers resiliency and scalability across the roles. These *VMs* are run on Microsoft Windows Server 2019 and get re-deployed during patches or updates.

Customers who are consuming services from a Microsoft Azure Stack Hub instance have access to a **Tenant Portal**, which allows them to select which services they wish to consume. This Tenant Portal is familiar to most customers as it is the same portal they would normally see if they were using Microsoft Azure in the public cloud. The cloud operators running the Microsoft Azure Stack Hub also have access to a portal, which is similar to the Azure portal, to perform operator tasks such as monitoring and updates. However, it offers greater functionality to allow them to create offerings to be exposed by Microsoft Azure Stack Hub. We will become more than familiar with this portal throughout this book.

For a traditional virtualization platform such as *Hyper-V*, tools such as *Hyper-V manager* would normally be used by IT administrators to create virtual machines and more. However, these are replaced in Microsoft Azure Stack Hub with the administration portal, as shown in the following screenshot:

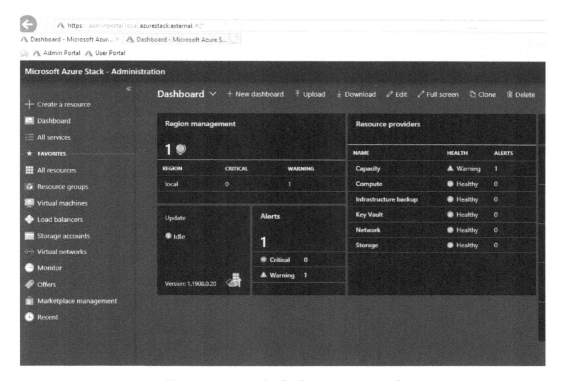

Figure 2.2 – Azure Stack administration portal

The preceding screenshot is an example of the **Microsoft Azure Stack Hub administration portal**, which would typically be found at `https://adminportal.local.azurestack.external`. This is where the cloud operator will spend most of their time while working with Microsoft Azure Stack Hub.

Next, we will look at the **infrastructure roles** and the roles they play.

Core infrastructure roles

The following diagram shows the core architecture of Azure Stack Hub:

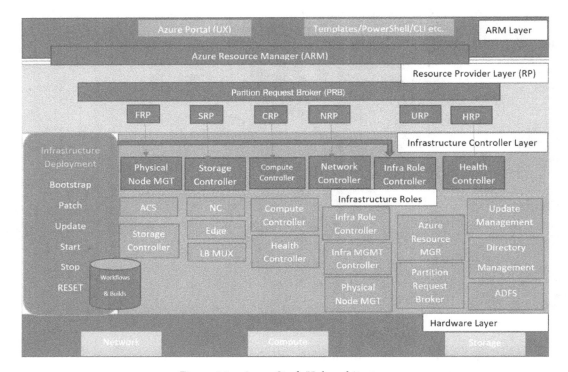

Figure 2.3 – Azure Stack Hub architecture

As you can see, Microsoft Azure Stack Hub has a set of predefined infrastructure roles included in the base architecture, and each has a particular role to play. These core roles include the following:

- **AzS-ACS01**: Azure Stack Hub Storage Services
- **AzS-ADFS01: Active Directory Federation Services (ADFS)**
- **AzS-CA01**: Certificate authority services for Azure Stack Hub role services
- **AzS-DC01**: Active Directory, DNS, and DHCP services for Azure Stack Hub
- **AzS-ERCS01**: Emergency Recovery Console VM

- **AzS-GWY01**: Edge gateway services such as VPN site-to-site connections for tenant networks

- **AzS-NC01**: Network Controller, which manages Azure Stack Hub network services

- **AzS-SLB01**: Load balancing multiplexer services in Azure Stack Hub for both tenants and Azure Stack Hub infrastructure services

- **AzS-SQL01**: Internal data store for Azure Stack Hub infrastructure roles

- **AzS-WAS01**: Azure Stack Hub administrative portal and Azure Resource Manager services

- **AzS-WASP01**: Azure Stack Hub user (tenant) portal and Azure Resource Manager services

- **AzS-XRP01**: Infrastructure management controller for Azure Stack Hub, including compute, network, and storage resource providers

Now, let's turn our thoughts to the **hardware life cycle host**.

Hardware life cycle host (HLH)

The **hardware life cycle host** or **HLH** is a physical server connected to the **BMC** network that is external to the Microsoft Azure Stack Hub environment and is part of the appliance from an OEM hardware vendor. It can be used to run a partner's life cycle management software. It is used for its hardware monitoring software, firmware configuration, and software updates. It can also be used for emergency management and any hardware troubleshooting. It is primarily used in the deployment process, where it is responsible for running the Azure Stack Hub's **Deployment Virtual Machine** (**DVM**) for the duration of the deployment. This server is connected to the *BMC* switch on the network and typically runs Microsoft Windows Server 2016/2019, either *Standard* or *Datacenter* edition. It has the Hyper-V role enabled and must meet the Microsoft Azure Stack Hub's security requirements. The HLH contains the generic Microsoft Azure Stack Hub image media, along with any associated OEM extension packages needed for the hardware. Not all OEM vendors use the HLH and for those that do, they are responsible for managing this server.

We will be coming back to the HLH again when we cover the *deployment process* later in this book but for now, we will move on to some **Azure Stack Hub concepts**.

Azure Stack Hub concepts

Scale units are a unit of capacity expansion and there are one or more scale units within a region. A scale unit is associated with a single region. There are 4 to 16 servers (at the time of writing) within a scale unit, and these would align with a hardware **SKU** from the OEM vendor. Each of these servers must have exactly the same characteristics (CPU, RAM, disk, and networking). The following diagram shows the typical setup for a scale unit with **servers**, **switches**, and the **HLH**:

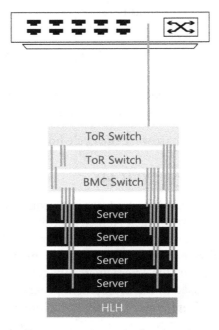

Figure 2.4 – A standard scale unit

Regions are a set of scale units that share the same physical location. The region would be controlled by one administration portal that's both physical and logical. The region should have a flat layer 3 network with high bandwidth and low latency. The scale units within a region would normally be the same type of hardware but may have different types of hardware from the same vendor. This is the vision from Microsoft and should be how this will be designed in the future, but at the time of writing, this is not currently possible. The following diagram shows how the scale units fit into a region, which is followed up by it being fitted into a private cloud:

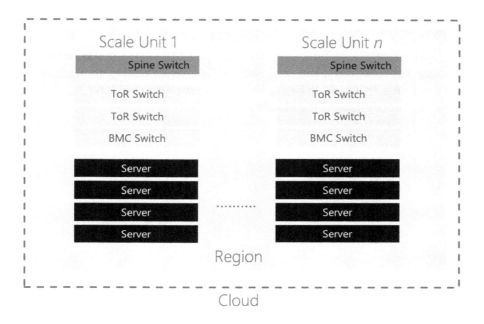

Figure 2.5 – Region and cloud ARM instances

A private cloud is a single instance of **Azure Resource Manager** (**ARM**), and one or more regions may be under the management of this ARM. A cloud supports different hardware vendors across multiple regions. It should be noted that this is very much a future state capability of Azure Stack Hub, but this is not available at the time of writing. As it stands today, each region is independent and would therefore need to be managed independently.

In this section, we learned about the different components of the Microsoft Azure Stack Hub, including the HLH. We will now move on and look at *capacity*, *scalability*, and *resiliency*.

Learning about capacity, scalability, and resiliency

Now that we understand the architecture that underpins the Microsoft Azure Stack Hub, let's turn to **capacity**, **scalability**, and **resiliency**. When we are talking about capacity, there are certain things that need to be considered and should act as inputs to the capacity planning process. These include the following:

- Servers
- Cores
- Memory
- GPU
- Storage devices (capacity)
- NIC performance
- Storage performance
- Infrastructure use (VMs and software updates)
- Form factors
- Application behaviors

When you are planning for a Microsoft Azure Stack Hub implementation, you need to consider the hardware configuration choices as these will have a direct impact on the overall capacity of the Microsoft Azure Stack Hub solution.

You need to think about choices with regards to *CPU*, *memory density*, GPU, *storage configuration*, network speed, and the overall solution's scale or number of servers. But the determination of usable capacity will be different to what it would have been with a traditional virtualization platform, since some of the capacity will already be in use due to the hyperconverged nature of Microsoft Azure Stack Hub.

Microsoft provides an Azure Stack Hub capacity planner, which is an Excel spreadsheet that is used to show how different allocations of computing resources would fit across different hardware settings. This capacity planner spreadsheet consists of the following sheets:

- *Version-Disclaimer*: This describes the purpose of the calculator, version number, and release date.

- *Instructions*: This provides step-by-step instructions on how to model capacity planning for a collection of **virtual machines** (**VMs**).

- *DefinedSolutionSKUs*: This is a table with up to five hardware definitions. The entries are only examples. These can be changed so that they match the system configurations being considered.

- *DefineByVMFootprint*: This allows the user to find the appropriate hardware SKU by comparing the configurations with different sizes and quantities of virtual machines.

- *DefineByWorkloadFootprint*: This allows the user to find the appropriate hardware SKU by creating a collection of Azure Stack Hub workloads.

Since Microsoft Azure Stack Hub is delivered as an integrated system that's installed and configured by the OEM hardware vendor, each of these vendors is likely to have their own capacity planning tools that can also be used alongside the **capacity planner** from Microsoft.

The **base scale unit**, which consists of at least four servers, can then be extended by adding a server node to the scale unit. Servers can be added individually to a scale unit, with up to a maximum of 16 nodes per scale unit being allowed. Once the maximum number of servers has been reached (16, in this case), then an additional scale unit would be required for us to continue scaling out the solution.

Each scale unit is designed and configured for resiliency to ensure that there is no single point of failure within the scale unit. We can look at each component individually to see where resiliency can be factored in. The following diagram shows the resiliency that's built into a scale unit:

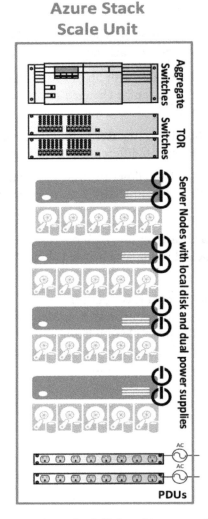

Figure 2.6 – Azure Stack Hub architecture resiliency

Let's take a look at how this resiliency is built across each of the layers:

- **Switch resiliency**: A single top of rack switch can fail, so a single scale unit will always include two top of rack switches to allow for redundancy.

- **Server resiliency**: A single server node can fail if disk failures are limited to one other server node. A third server node down or failed disks in a third server node will take the affected virtual disks offline.

- **Disk resiliency**: A single disk can fail anywhere in the scale unit. Multiple disks can fail, provided that they are limited to two server nodes.

- **PDU resiliency**: A single PDU or power supply can fail with no issues.

- **Network resiliency** : A single network card can fail, and traffic can still be routed as each server consists of two network cards.

We will now walk through each component in the hyperconverged infrastructure in turn, starting with the *compute layer*.

Understanding compute

The **compute architecture** is supported by the servers in the scale unit that are running Windows Server 2019 Standard or Datacenter edition. These Windows Servers then form *Hyper-V clusters*. Sitting on top of these Hyper-V clusters are the **Compute Resource Provider** and **the Azure Guest Agent**. Running on the Compute Resource Provider are the VMs, VM Extensions, and Platform Images, which can all be controlled from the **Azure Resource Manager** (**ARM**), as shown in the following diagram:

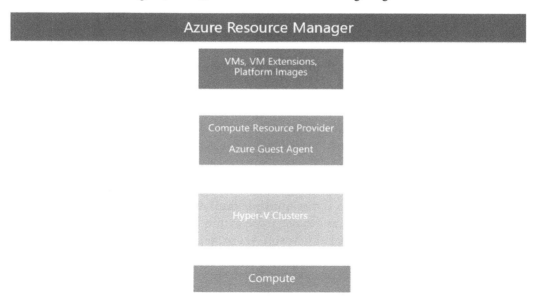

Figure 2.7 – The Azure Stack Hub compute architecture

The *Compute Resource Provider* allows Microsoft Azure Stack Hub tenants to create their own virtual machines. This includes the ability to create virtual machine extensions along with the virtual machines. The orchestrator component is responsible for locating the correct host to run the VM on. The virtual machine extension service assists with providing *IaaS* capabilities for *Windows* and *Linux* machines. For example, you can use the Compute Resource Provider to provision a Linux virtual machine and run **Bash** scripts during deployment to configure the virtual machine. In a similar vein, it is possible to use the Compute Resource Provider to provision a Windows machine and then run **SysPrep** to configure the virtual machine.

Microsoft Azure Stack Hub supports a subset of VM sizes compared to the Azure public cloud. At the time of writing, Microsoft Azure Stack Hub currently supports *A*, *D*, *DS*, *F*, and *N* virtual machine types of various sizes. The A virtual machine type is the general entry-level machine most suited for an office worker role. The D and DS virtual machine types are used for general workloads, while the F virtual machine types are better suited for compute-intensive workloads.

Emulation is imperfect and is focused on uniform in-guest quantities. RAM, CPU core count, and disk size quantities are aligned with Microsoft Azure. CPU performance is better than the equivalent in Microsoft Azure for low-load environments, depending on the vCPU:Core ratio. CPU performance is non-deterministic for a given VM size, so it is highly probable that an A VM size image will run the same as a Dv2 VM image.

There is a limit on the total number of virtual machines that can be created, and this is to avoid solution instability. At the time of writing, this limit was set to 60 virtual machines per server with a total solution limit of 700 virtual machines, depending on the number of servers. However, this includes the 30 infrastructure role virtual machines.

Microsoft Azure Stack Hub utilizes a placement engine, which determines where virtual machines are to be placed over the available compute nodes. We will revisit this in *Chapter 11, Grasping Storage and Compute Fundamentals,* later in this book, when we take a closer look into the *compute features* available in Microsoft Azure Stack Hub.

To be able to achieve high availability of a multi-virtual machine production within Microsoft Azure Stack Hub, virtual machines are incorporated into an availability set, which spreads the virtual machines across **multiple fault domains**. A fault domain within an availability set is defined as a single node within a scale unit, while in Azure, the fault domain would be an entire rack. Microsoft Azure Stack Hub allows an availability set to have a maximum of three fault domains, so in that way it remains consistent with Microsoft Azure. Virtual machines that are placed within an availability set will be physically isolated from each other. This is done by spreading them as evenly as possible throughout multiple fault domains.

Virtual machine scale sets use availability sets in the backend and ensure that each virtual machine scale set instance is placed in a different fault domain. That is to say that they use different Microsoft Azure Stack Hub infrastructure nodes. A virtual machine scale set is a compute resource within Azure Stack Hub that is used to deploy and manage a set of identical virtual machines.

Microsoft Azure Stack Hub is designed to keep virtual machines running that have been successfully provisioned. As an example, if a host is taken offline due to a hardware failure, Azure Stack Hub attempts to restart the virtual machine on another host. Another example comes from the patching and updating process for the Microsoft Azure Stack Hub software, where the workload will be drained from a node and moved to another node within the solution while the original node is updated.

Managing or moving virtual machines can only be achieved when there is reserved memory capacity to allow the restart or migration to happen. As a result, part of the total host memory is reserved, which means it is unavailable for tenant virtual machine placement. In terms of capacity, this equates to one node, which allows one node to be removed from the cluster for maintenance purposes.

When using the administration portal of Microsoft Azure Stack Hub to look at your free and used memory, it is worth noting that used memory is made up of several components. The components that consume memory within a Microsoft Azure Stack Hub solution include the following:

- **Host OS usage** or **reserve**: The memory that's used by the **operating system** (**OS**) on the host, virtual memory page tables, processes that are running on the host OS, and the Spaces Direct memory cache. Since this value is dependent on the memory used by different Hyper-V processes running on the host, it can fluctuate over time.

- **Infrastructure Services**: The infrastructure virtual machines that make up Azure Stack Hub (31 virtual machines taking up 242 GB).

- **Resiliency reserve**: Azure Stack Hub reserves a portion of the memory to allow for tenant availability during a single host failure, as well as during patches and updates to allow for virtual machines to be live migrated successfully.

- **Tenant virtual machines**: These are the tenant virtual machines that are created by Azure Stack Hub users. In addition to running virtual machines, memory is consumed by any virtual machine that has landed on the fabric. Any virtual machine that is in the state of creating or failed and any virtual machine that is shut down on the host is still consuming memory.

- **Value-add resource providers**: Virtual machines deployed for the value-add resource providers such as *SQL*, *MySQL*, *AppService*, and so on.

The calculation results that are received from the Microsoft capacity planner tool will be the total available memory that can be used for tenant virtual machine placement. This memory capacity would be for the entire Microsoft Azure Stack Hub scale unit.

The other calculation that needs to be carefully considered for planning purposes is the CPU core. Careful consideration of the SKU should take into account the workload that you intend to host. This is because different types of virtual machines will have different memory requirements. For example, the F virtual machine type has a RAM:CPU ratio of 2 GB RAM per vCPU, while a D virtual machine has a ratio of 4 GB RAM per vCPU. This means that if the calculation of the vCPU to CPU core ratio is incorrect, then there is the potential to either run out of cores or run out of memory.

We will cover compute in greater detail later, when we walk through the features of Microsoft Azure Stack Hub, but for now, we just need to understand how compute is scalable, as well as the components that underpin Microsoft Azure Stack Hub. Next, we are going to turn our thoughts to the networking component of the Microsoft Azure Stack Hub.

Getting to know the networking component

In a similar vein to compute, **networking** is supported by the servers, and the *vSwitch gateway* sits above them on the data plane. The **Network Resource Provider** sits above this, supporting the network controller and the software load balancer. Above this is the *vNICs*, *Gateways*, *vNets*, and more, all of which all administered using the *Azure Resource Manager*, as shown in the following diagram:

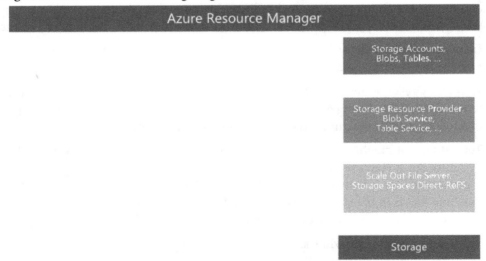

Figure 2.8 – The Azure Stack Hub networking architecture

From a *physical network switching* topology point of view, there are two connectivity types supported within Microsoft Azure Stack Hub. **Data switching network** connectivity is used for the Windows Server 2019 converged NIC, software-defined networking, and storage. **BMC switching network** connectivity is used for physical host control and any third-party hardware monitoring that's required by the OEM vendor.

Resiliency is built into the networking through dual *top of rack switches* initially and then it is expanded at the data layer with dual ports on a *single physical NIC* with a minimum 10 GbE. It also makes use of switch-embedded teaming, which is a feature of Windows 2016/2019 for port/link resiliency.

Within a scale unit, each server is connected to the BMC switch and both of the *top of rack* switches. The two *top of rack* switches are connected to each other via a **multi-chassis link aggregation group** (**MLAG**). The two *top of rack* switches are also connected to the BMC switch for switch management. BMC is then also used for out-of-band server management.

When more than one scale unit is used, an aggregate switch is required for each scale unit. Each of the *top of rack* switches in each scale unit is connected to each aggregate switch. The aggregate switches are then connected to the customer's core network backbone.

The *top of rack* switch configuration is pre-configured during the deployment process, and it expects a minimum of one connection between the *top of rack* switch and the border when using BGP routing, and then a minimum of two connections (one per *top of rack* switch) between the *top of rack* switch and the border when using static routing. You need a maximum of four connections for either routing option.

Microsoft Azure Stack Hub is built on Windows Server 2019 failover clustering and storage spaces' direct technologies. Because of this, some of the Microsoft Azure Stack Hub's physical network configuration is done to utilize traffic separation and bandwidth guarantees. This ensures that the storage spaces' direct communications can meet the performance and scale required for the solution.

Microsoft Azure Stack Hub benefits greatly from many **software-defined networking** capabilities in Windows Server 2019, including the following:

- **Network Controller**:
  - Central control plane
  - Fault-tolerant

- **Virtual Networking**:

  - BYO address space

  - Multiple subnets

  - Distributed router

- **Network Security**:

  - Distributed firewall

  - Network security groups

  - BYO virtual appliances

- **Robust Gateways**:

  - Robust availability model

  - Multi-tenancy for all modes of operation

- **Software Load Balancing**:

  - L4 load balancing

  - NAT for tenants and Azure Stack Hub infrastructure

- **Data Plane Improvements**:

  - Performance: 10 GB, 40 GB and higher

  - RDMA over virtual switch

This rich software-defined network ecosystem is all configured automatically during the deployment process of Microsoft Azure Stack Hub and can be used immediately, as soon as the deployment is completed.

The Microsoft Azure Stack Hub network is architected to use five subnets for the following purposes:

- Switch infrastructure

- BMC (iLO, DRAC, and so on)

- Infrastructure

- Private

- Public VIP

Microsoft Azure Stack Hub uses an IP address for the following services:

- Tenant Portal

- Admin Portal

- Azure Resource Manager

- Storage (Blob, Table, Queue)

- xRP, Key Vault

- ADFS, Graph

- Site-to-Site Endpoint

- Tenant 1, Tenant 2 … Tenant N

The logical network that's configured within Microsoft Azure Stack Hub can be seen in the following diagram:

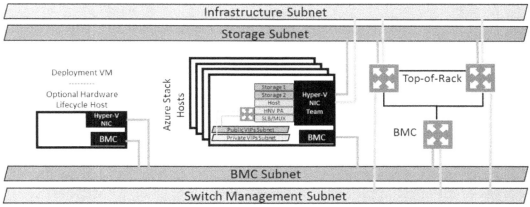

Figure 2.9 – The logical network architecture for Azure Stack Hub

We will return to networking again in a later chapter, but at this point, we are satisfied that we have an understanding of the components involved in networking for Microsoft Azure Stack Hub. In the next section, we shall look at the last component of the *Hyperconverged Infrastructure Storage*.

Introducing storage

As with both *compute* and *networking*, **storage** is supported by the servers and the storage spaces directly via **ReFS**. Above this is the storage resource provider and above this is the storage account. As with the others, this is all controlled by the Azure Storage Resource Provider, as shown in the following diagram:

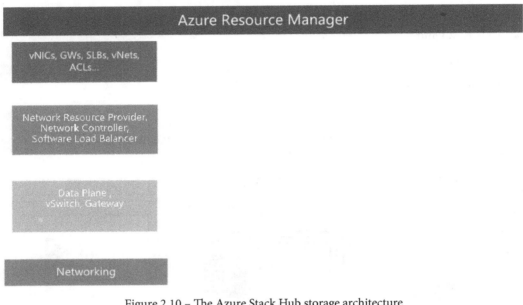

Figure 2.10 – The Azure Stack Hub storage architecture

The Microsoft Azure Stack Hub storage stack consists of the following components:

- Storage Spaces
- Software Storage Bus
- Servers with local disks

Microsoft **Storage Spaces Direct** provides a single scalable pool with all disk devices except for the boot drive. Multiple virtual disks are created per pool and mirrored across the servers.

The **Software Storage Bus** provides the storage bus cache and leverages SMB3 and SMB Direct.

The servers can be provisioned with SATA, SAS, or NVMe disks, depending on the vendor configuration that's selected before deployment.

The Microsoft Azure Stack Hub storage stack is shown in the following diagram:

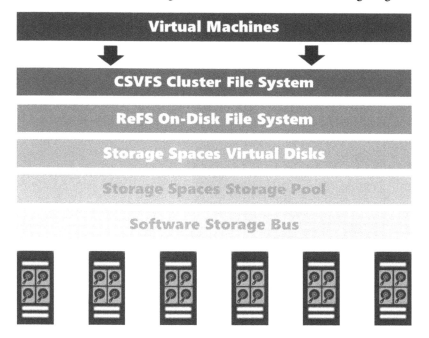

Figure 2.11 – The Azure Stack Hub storage stack

Cache is an important concept within Microsoft Azure Stack Hub and is an integral part of the Software Storage Bus. The cache is scoped to a local machine. The *storage cache* is agnostic to both the storage pool and the associated virtual disks. Storage caching is configured automatically, if needed, when Storage Spaces Direct is enabled and the cache behavior is dependent on the type of operation that is being performed. The following diagram shows the drives that are used for caching and capacity in an Azure Stack Hub scale unit:

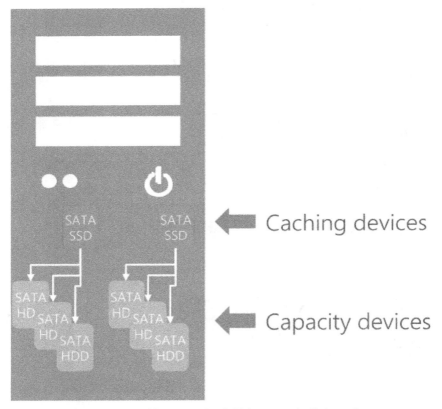

Figure 2.12 – The Azure Stack Hub storage built-in cache

Azure data volumes have resiliency granted from multiple data copies and, like Azure, data has three copies in Microsoft Azure Stack Hub due to it leveraging a three-way mirror. This means that normal mirror sets can survive the loss of a single physical disk, while a Microsoft Azure Stack Hub three-way mirror can survive the loss of two physical disks. The volume type of the mirror is optimized for performance based on the *ReFS filesystem*.

Microsoft Azure Stack Hub delivers Azure-consistent *blobs*, *tables*, and *storage accounts*. It provides comprehensive private cloud service provider manageability. It can be deployed in either enterprise-private clouds or hosted public clouds from service providers. It supports *IaaS storage* (page blobs) and *PaaS storage* (block blobs and tables). It is highly reliable, resilient, and scalable cloud storage based on standard hardware.

In Storage Spaces Direct for Microsoft Azure Stack Hub, the default is to keep three copies of the data split between three server nodes. Microsoft Azure Stack Hub VMs are clustered and split across three nodes. Three copies of all the data that's been written to disk are mirrored in the same manner. This means you can lose a whole server due to failure or during patch and update and still have two-copy redundancy.

There is also storage capacity, which is used for the OS, local logging, dumps, and other temporary infrastructure storage needs. This local storage capacity is kept separate from the storage devices, which are managed by the Storage Spaces Direct configuration. As we mentioned previously, the remainder of the storage devices are placed in a single pool of storage capacity. This is done regardless of how many servers are included in the scale unit.

The Microsoft Azure Stack Hub infrastructure manages all the storage capacity and is also responsible for directly allocating this storage capacity. As we mentioned previously, Microsoft Azure Stack Hub automates the design and allocation of this storage capacity when installing and enabling Storage Spaces Direct. The Microsoft Azure Stack Hub considers the resiliency, reserved capacity for rebuilds, and other details to be part of the automated design.

When a Microsoft Azure Stack Hub appliance has only one drive type, then all the drives are used for capacity. However, if the configuration contains two types of disks (SSD and HDD, NVME SSD and SAS SSD, and so on), then the fastest drive types are automatically used for caching.

This caching behavior is determined automatically by Storage Spaces Direct, depending on the kinds of drives in the configuration. If caching is enabled for SSDs, then only write operations are cached. This is because this reduces the wear on the capacity drives, which, in turn, reduces the cumulative traffic to the capacity drives and therefore extends their life.

Important Note

The Microsoft Azure Stack Hub appliance can be configured as a hybrid deployment with both HDD and SSD drives. The faster drive types will be used for the cache drives, while the remainder of the drives will be used as capacity in a single pool. Tenant data would be placed on a capacity drive. There is no guarantee that selecting premium disks or selecting a premium storage account type will result in the objects being allocated to these faster drives.

The available storage is partitioned into separate volumes by the storage service and are allocated to hold both tenant and system data. These volumes combine the drives in the storage pool to provide the fault tolerance, scalability, and performance benefits of Storage Spaces Direct.

Within the Microsoft Azure Stack Hub **Storage Pool**, there are three types of volumes that store different datasets. They are as follows:

- *Infrastructure*: Hosts files used by Azure Stack Hub infrastructure VMs and core services.

- *VM Temp*: Hosts temporary disks attached to tenant VMs, and that data is stored here.

- *Object Store*: Hosts tenant data servicing blobs, tables, queues, and VM disks.

For a multi-node deployment, three infrastructure volumes will be created automatically. For a base, four-node deployment of Microsoft Azure Stack Hub, then four equal VM temp and four equal object store volumes are created. If a new node is added to the solution, then an additional VM temp and object store volume will be created.

Summary

In this chapter, we looked at the different components of the underlying architecture for Microsoft Azure Stack Hub. We learned that the hyper-converged infrastructure consists of compute, storage, and networking components in a single appliance. We should now be able to describe the Microsoft Azure Stack Hub storage stack. We should also be able to describe the compute stack, alongside the networking stack for Microsoft Azure Stack Hub.

With that, we have been introduced to the different roles involved in administering an Azure Stack Hub instance and had our first introduction to the Azure Stack Hub Administration portal. We should now be familiar with the two different roles that are available: *Cloud Operator* and *Cloud Architect*.

We then looked at how to configure for capacity, resiliency, and performance in Microsoft Azure Stack Hub. We looked at the considerations we should take into account when we're planning to implement Microsoft Azure Stack Hub.

Finally, we were introduced to the HLH and the concepts of scale units, regions, and the private cloud.

We will be revisiting the HLH again in the next chapter, when we look at its deployment process in more detail.

3
Azure Stack Hub Deployment

The final chapter in this first part of the book goes through the **deployment** process in detail, including the prerequisites. This chapter details the information required and deployment options that need to be taken upfront before deployment can commence. We will then cover **post-deployment** and **troubleshooting tips** for the deployment process.

By the end of this chapter, you should have a complete understanding of the deployment process. This will be beneficial for those of you looking at taking the *AZ-600* exam as there are likely to be some questions that relate to this subject in the exam.

In this chapter, we're going to cover the following main topics:

- Understanding the cloud operating model
- Overviewing data center integration in depth
- Reviewing infrastructure and cloud service management
- Preparing for deployment
- Getting to know HLH and DVP deployment
- Deploying Azure Stack Hub
- Post-deployment and troubleshooting

Technical requirements

You can view this chapter's code in action here: `https://bit.ly/3zfMYnn`

Understanding the cloud operating model

The Azure Stack Hub cloud operating model mirrors the cloud operating model that underpins the Azure public cloud. Microsoft is the operator of the Azure cloud and provides services to their customers. Microsoft employs a vast number of engineers to maintain and operate this cloud environment. With Azure Stack Hub, the organization that operates the Azure Stack Hub effectively takes on the role of Microsoft. They provide the services to their end users and employ engineers to maintain this infrastructure.

The term operating model has many definitions but there are certain terms that are consistent across different operating models:

- **Business model**: Business models are normally defined by organizations and cover corporate values or mission statements. The business model as a minimum should convey the what and the why based on forms of financial projections. For business models to be effective, they should include high-level statements that provide directional goals for the organization. These goals can then be represented through the use of KPIs and metrics.

- **Customer experience**: Business models look at how the organization interacts with their customers and part of the why needs to cover the customer experience. Most successful organizations realize that the experience of their customers is key to their own success.

- **Digital transformation**: Digital transformation has been around for years as a buzzword in the industry. It is a vital component for fulfilling modern business models. In order to deliver business value, the customer experience must take into account the digital experience. This is the process of digital transformation.

- **Operating model**: The operating model covers the how and the who for operating the business. The operating model defines how people work together to accomplish the goals defined in the business model.

- **Cloud adoption**: Cloud adoption is a key part of the operating model. This is a strong enabler for delivering the right technologies and processes needed to successfully deliver on the modern operating models.

With an understanding of these common terms, we can now take a look at the definition of the cloud operating model for Azure Stack Hub.

Defining the cloud operating model

Before deploying Azure Stack Hub for your cloud architecture, it is important to think about how you want to operate your cloud. This means thinking about your strategic direction and defining the future state of your cloud operating model.

Before cloud technologies existed, technology teams established operating models to define how technology would be used to support the business. Any IT operating model includes several factors, such as business strategy, change management, organization, and operations. Each factor has an important part to play in the long-term operation of the technology.

As technology moves toward the cloud, the operating processes need to adapt. A cloud operating model is essentially a collection of processes and procedures that define how you want to operate technology in the cloud. This equally applies to operating the technology in Azure Stack Hub, which is more or less your own private cloud.

The purpose of a cloud operating model is to think about a higher level of operations and focus on digital assets and workloads. The cloud operating model shifts the thinking away from just keeping the lights on to ensuring that workloads are consistent.

Microsoft has tools available to help you to define your cloud operating model. One such tool is the **Cloud Adoption Framework**. The Cloud Adoption Framework helps to break down the different aspects of the operating model into methodologies. The Cloud Adoption Framework is shown in the following diagram:

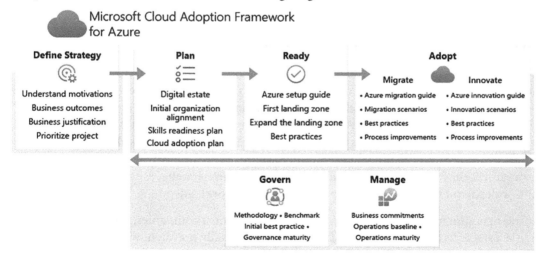

Figure 3.1 – Microsoft Cloud Adoption Framework

This framework, although originally designed for customers looking to adopt public Azure, is equally applicable to Azure Stack Hub.

Certain areas of the Cloud Adoption Framework are designed to be incremental methodologies to help grow the operating model over time:

- **Manage**: Aligns the ongoing processes for the operational management of the technology
- **Govern**: Aligns to the governance and compliance requirements

Your cloud operating model should represent how you want your organization to operate with cloud technology.

As you define the operating model for your organization, the environmental readiness should match your operations, governance, security, and organizational requirements.

The business strategy and cloud adoption plans are both inputs that need to be considered when defining your cloud operating model.

Operating models are specific to the business requirements that they are designed to support. That being said, there are still some common patterns that can be followed based on the current requirements and constraints. Let's take a look at some of these common operating model patterns now.

Common operating model patterns

Common operating models can be based on the range of complexity, from the least complex (decentralized) to the most complex (global operations).

Operations are always complex, so by limiting the scope of operations to a single workload or a small collection of workloads, we can control this complexity. A decentralized operational model is the least complex of the common operating models as with this form of operations, all workloads are operated independently by dedicated teams. There are some advantages to this type of model, such as cost management for a single workload and the use of automation to standardize for a single workload, but as the organization grows, this model can soon become cumbersome, especially when other units or departments need to run the same or similar workloads.

Central operations tend to be the normal approach for most companies, especially when there is a stable workload. This operating model is especially prevalent when companies make use of **commercial off-the-shelf** (**COTS**) products across their estate. This operating model has advantages over the decentralized operating model as there are economies of scale and standardized operations. This model is ideal for cloud-based operations as it allows tight control of integration touchpoints and also allows the organization to control costs.

The enterprise operating model builds on the centralized model but replaces the central IT with a more facilitative cloud operations team. This operating model becomes more about working with the business units to allow them to have more control in the decision-making process, rather than enforcing decisions on them as in the previous model. There is still the ability to control costs, but these can be drilled down to individual departments, which become more autonomous.

The final common operating model is the distributed operating model. This model is really for organizations where their existing operating model is perhaps too tightly ingrained in the organization to allow the whole organization to move to a new operating model. For global organizations, there may well be various compliance requirements that might prevent certain business units from changing their operating model. For these organizations, distributing their operations is the only option and this is by far the most complex of the common operating models. This model is not really suitable for an Azure Stack Hub operating model without running multiple Azure Stack Hub instances in different regions. This becomes harder to control both from a cost basis and with regards to standardization on products or workloads.

Now that we have an understanding of the different cloud operating models, we can move on to take a look at the integration into the data center in the next section.

Overviewing data center integration in depth

In this section, we will take an in-depth look at the touchpoints for data center integration. You decide how Azure Stack Hub is going to be integrated into your data center but in reality, the integration is a collaborative project between you, the **original equipment manufacturer** (OEM) solution provider, and Microsoft. As we touched on earlier in this chapter when looking at deployment, some of the integration information is provided when completing the deployment worksheet with the OEM vendor. Decisions would have already been made about the following prior to deployment:

- The connection model:
  - Connected
  - Disconnected
- The identity model:
  - **Azure Active Directory (AAD)**
  - **Active Directory Federation Services (ADFS)**

- The licensing model:

  - Capacity-based billing

  - Pay-as-you-use billing

- Network integration

- Firewall integration

- Certificates

All of this information is gathered before deployment and incorporated into the deployment worksheet, which the OEM vendor will then use to perform the deployment of Azure Stack Hub. The following diagram shows Azure Stack Hub at the center of these decision points:

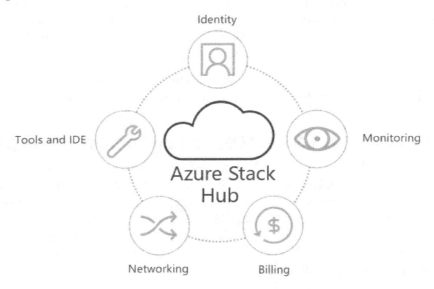

Figure 3.2 – Data center integration decision points

While researching for information that needs to be included in the deployment worksheet, you may find that you need to make pre-deployment changes to your networking environment. This can include reserving the IP address spaces for Azure Stack Hub or configuring switches, routers, or firewalls to prepare for the connectivity of Azure Stack Hub switches.

Azure Stack Hub is an integrated sealed system, which means the infrastructure is locked down both from a permissions and a network perspective. Network **access control lists (ACLs)** are used to block all unauthorized incoming traffic and all unnecessary communications between infrastructure components, as we will discuss in *Chapter 5, Securing Your Azure Stack Hub Instance*. Unlike most systems, there is no unrestricted administration access in Azure Stack Hub and all management and operations are conducted through the Azure Stack Hub administration portal or **Azure Resource Manager (ARM)**.

You will need to think about how you wish to name your Azure Stack Hub instance, including the namespace, region name, and external domain name. The external domain name of your Azure Stack Hub instance will be used for public-facing endpoints and will be in the `<region>.<external domain name>` format. The region name chosen for the Azure Stack Hub instance must be unique.

You must also choose a single specific time server that will be used to synchronize Azure Stack Hub. This time synchronization is critical to Azure Stack Hub and the infrastructure roles as it is used to generate the **Kerberos** tickets used to authenticate internal services to each other, which we will discuss later in *Chapter 5, Securing Your Azure Stack Hub Instance*. For the time server, you must specify an IP address as even though most components can resolve a **fully qualified domain name (FQDN)**, there are some components that can only support IP addresses. If Azure Stack Hub is deployed in the disconnected scenario, then the time synchronization server must be on the corporate network and accessible to the Azure Stack Hub instance.

When Azure Stack Hub is deployed in the connected scenario and used for hybrid cloud scenarios, then the connectivity to Azure also needs to be planned. There are two ways of connecting Azure Stack Hub **virtual networks (VNets)** to VNets in Azure that are supported by Microsoft. The first of these options is to use a site-to-site **virtual private network (VPN)**. This is a VPN connection that requires a VPN device or **Routing and Remote Access Service (RRAS)**. This creates a tunnel and all communication over this tunnel is encrypted and secure.

For hybrid scenarios, consideration must be made for the kind of deployment that the organization wants to offer and where it will be deployed. Considerations need to be made for whether there is the need to isolate network traffic per tenant and whether an intranet or internet deployment is required.

A single-tenant deployment of Azure Stack Hub means that from a networking standpoint, it is seen as a single tenant. There can be multiple tenant subscriptions but as with a standard intranet or **SharePoint** implementation, all network traffic utilizes the same network.

An Azure Stack Hub instance that is deployed as multi-tenant will require each tenant's network traffic to be isolated from other tenant traffic as each tenant subscription will be connected to different external networks outside of Azure Stack Hub.

An Azure Stack Hub instance that is deployed as an intranet will typically be deployed to a corporate network in a private IP address space with protection provided by firewalls. The public IP addresses are not true public IP addresses as they cannot be routed directly over the public internet.

Azure Stack Hub can also be deployed when connected to the public internet and utilize internet-routable public IP addresses for the public VIP range. This deployment will still sit behind a firewall for protection, but the public VIP range is directly accessible from the public internet and Azure.

The other option for connectivity for Azure Stack Hub is the use of **ExpressRoute**. ExpressRoute is a circuit that is provisioned by a provider such as Vodafone and is suitable for both the single-tenant or multi-tenant scenarios detailed here. The following diagram shows ExpressRoute connectivity for a single-tenant deployment of Azure Stack Hub:

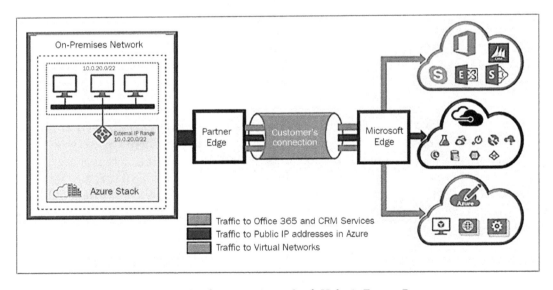

Figure 3.3 – Single-tenant Azure Stack Hub via ExpressRoute

In the previous diagram, ExpressRoute is shown as the **Customer's connection**.

In a similar vein, the following diagram shows the ExpressRoute connectivity for the multi-tenant scenario:

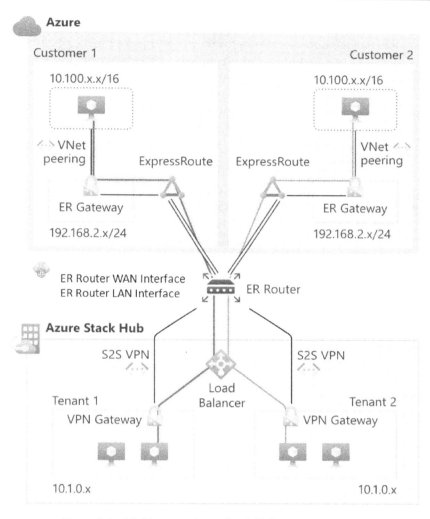

Figure 3.4 – Multi-tenant Azure Stack Hub via ExpressRoute

Each tenant creates a site-to-site VPN connection between their Azure Stack Hub VNet and the ExpressRoute router of the ExpressRoute provider. VNets from Azure are connected to the ExpressRoute router. The ExpressRoute router LAN interface requires one IP address for each Azure Stack Hub VNet.

It is possible to get a single view of all alerts from Azure Stack Hub by integrating into existing IT service management workflows for tickets or integrating into external data center monitoring solutions. Azure Stack Hub includes a **hardware life cycle host (HLH)**, which is a server that sits outside of Azure Stack Hub. Most OEM vendors will provide management tools on the HLH for hardware alerts, which can be extended to include alerts from Azure Stack Hub. For example, Lenovo installs an application called **XClarity** onto the HLH, which includes an administration portal. They also have XClarity Integrator for Azure Stack Hub, which allows the alerts generated by Azure Stack Hub to be visible within the XClarity admin portal.

There are plugins available for different monitoring solutions, including **Nagios** and **Microsoft System Center Operations Manager**, alongside the monitoring tools provided by the OEM vendor, and this list is likely to grow over time as the Azure Stack ecosystem continues to mature. Each of the tools that can be integrated for monitoring all share the same requirements. They must all be agentless as Azure Stack Hub is a sealed system, and no third-party agents can be installed.

It is also possible to integrate security logging into Azure Stack Hub. The following logs are collected:

- Windows system logs
- Windows application logs
- Windows Security Log
- Windows event logs
- Active Directory application logs
- Active Directory diagnostic logs
- DNS application logs
- DNS diagnostic logs

These logs are all collected by the **Azure Monitor agent** and are stored in the **fabric resource provider** (**FRP**) storage account and component storage account. They can be viewed through the use of Storage Explorer.

The following diagram shows the different touchpoints for data center integration that we will cover in this chapter:

Figure 3.5 – Data center integration touchpoints for Azure Stack Hub

In the preceding diagram, the red boxes are internal to Azure Stack Hub while the green boxes are external to Azure Stack Hub. The items marked with an asterisk represent integrations that require additional configuration.

We will be covering both certificates and network integration in more detail later in this chapter and in other chapters later in this book, but for now, we will move on from integration to take a look at cloud service management.

Reviewing infrastructure and cloud service management

In a typical infrastructure today, there are generally many silos of technology and responsibilities. Take the following diagram as an example of this:

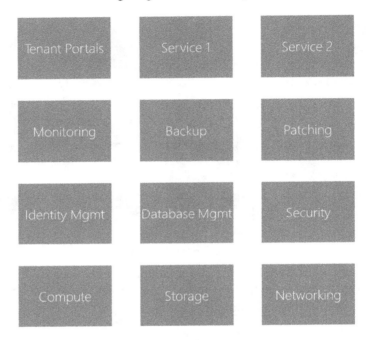

Figure 3.6 – Example of typical infrastructure today

This provides multiple points of infrastructure management with different UI or API connectivity. Each silo of technology is likely to require in-depth knowledge to manage and maintain. This adds greater complexity to IT management and means it is not easily administered by a single administrator or team.

Compare this to the cloud administration of Azure Stack Hub in the following diagram:

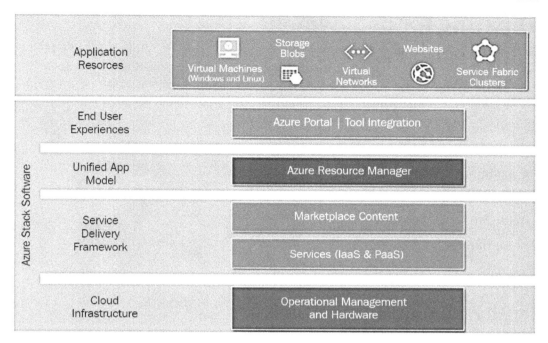

Figure 3.7 – Azure Stack Hub cloud administration visualized

The vision from Microsoft for cloud administration in Azure Stack Hub is to remove this complexity of infrastructure management with a single pane of glass. This single access point via a UI allows cloud administrators to operate and troubleshoot across all the infrastructure. This removes the need for multiple levels of deep knowledge for each silo as there are fewer knobs to turn. From an organizational standpoint, it then means that the IT department can concentrate on service levels and commitment to the business, rather than managing multiple silos of infrastructure.

This short section has shown that cloud service management is less complex than the traditional infrastructure management at play in most organizations today. The next section will cover the details of some key decisions that need to be taken prior to deployment.

Preparing for deployment

Before deploying Azure Stack Hub into an environment, certain planning needs to take place, including the touchpoints required for integration with the existing data center. Your OEM vendor will work with you throughout the planning stage to ensure everything is ready prior to the deployment. Some of the areas that you need to consider when planning for Azure Stack Hub include the following:

- Connection model
- Billing
- Networking
- Identity
- Certificates

We will look at these crucial decisions first before walking through the deployment.

Identity prerequisites

One of the key decision points before implementing Azure Stack Hub is choosing whether the Azure Stack Hub deployment is connected or not connected to Azure. This is a one-time decision that must be made at deployment time and it cannot be changed later without a re-deployment of the Azure Stack Hub solution.

There are a couple of key differences between the connected and disconnected scenarios as connected does give you more flexibility. In the connected state, it is possible to use AAD or ADFS, while in the disconnected model, the only option is to select ADFS.

> **Important note**
> If ADFS is used for identity management, then even in a connected scenario multi-tenancy is not supported.

The other main difference between the two scenarios is that from a billing standpoint, disconnected only supports a capacity-based billing model, while in the connected scenario, you have the choice of either a capacity-based or consumption-based billing model.

Connected scenario prerequisites

For the connected scenario, as mentioned, you have a choice of identity providers and these have implications for the functionality available from the platform.

ADFS can be chosen as an identity management store, which means the local organization's Active Directory will be used. This restricts the platform in the respect that multi-tenancy is not supported. If ADFS is selected as the identity provider, then the Graph service will be used.

When AAD is selected as the identity management store, then there are two AAD accounts required. The first one is used to create applications and service principals for infrastructure services. The second account is used to establish the billing relationship with the Azure commerce backend.

> **Important note**
> The two accounts can be the same account or can be completely different accounts.

Disconnected scenario prerequisites

In the *disconnected* scenario, you are constrained to a single identity source selection with ADFS. The disconnected scenario also requires the **Graph** service. The Graph service requires a service account with read-only access to the target Active Directory forest. The Graph service can only communicate with a single **Active Directory Domain Services (ADDS)** forest. The Graph service also requires access to an ADDS global catalog server.

Disconnected scenario limitations

In the disconnected scenario, one of the limitations is that due to the selection of ADFS as the identity management, multi-tenancy is not supported. It also means that as it is not connected to Azure, some features may be impaired or unavailable.

> **Important note**
> Choosing **Do not connect to Azure** does not strictly mean that you cannot connect the Azure Stack Hub instance to Azure for hybrid scenarios for tenant workloads.

The other limitation in a disconnected scenario is that marketplace syndication is not supported.

Registration with Azure is still required in a disconnected scenario but it is not automatic as it is in the connected scenario and will require the use of a separate connected device.

It is crucial that the decision around connected or disconnected is made upfront before deployment begins as this is a one-time decision that cannot be changed without redeployment of the whole platform.

Licensing model

You must think about which licensing model you want to use. The licensing model you can choose from is dependent on the connection model that has been chosen. If Azure Stack Hub is connected to the internet, then you can choose between pay-as-you-use or capacity-based licensing. If Azure Stack Hub is disconnected, then only capacity-based licensing is allowed, which is due to the fact that pay-as-you-use licensing requires connectivity to Azure to be able to report usage.

Networking prerequisites

Before deploying Azure Stack Hub, there are some prerequisites that need to be considered for both the physical and logical networks. Some of the decisions made about the network at this point cannot be changed post-deployment without the need for a redeployment, such as subnets.

Physical networking

Azure Stack Hub is reliant on a resilient and highly available physical network infrastructure to support the operations and services. To be able to integrate Azure Stack Hub into the data center network, it uses uplinks from the **top-of-rack** (**TOR**) switches to the closest switch or router, which is also referred to as the border. The TORs can be uplinked to a single border or a pair of borders. The TOR is pre-configured as part of the deployment based on settings taken from the deployment worksheet we discussed earlier. It expects at least one connection between the TOR and border when using **Border Gateway Protocol** (**BGP**) routing and at least two connections between the TOR and border when using static routing. Either routing option will support up to a maximum of four connections. These connections must have at least 1 GB speed and are restricted to SFP+ or SFP28 media. Microsoft recommends the use of BGP over static routing.

The connection outbound from Azure Stack Hub to the border switches are Layer 3 connections and each has its own /30 subnet, shared with the connecting border switch port. This is not a Layer 2 connection. For example, the IPv4 settings could be 10.128.0.28/30 and 10.128.32/30. The use of BGP requires that you set a 16-bit **autonomous system number** (**ASN**), public or private.

Static routing would require more manual configuration as each border device must be configured with static routes that point to each one of the four P2P IPs set between the TOR and the border device. A default static route is configured in the TOR to send all traffic to the border devices. TOR switches are configured with static route 0.0.0.0/0 to border the P2P address.

The next diagram shows the recommended physical network topology from Microsoft:

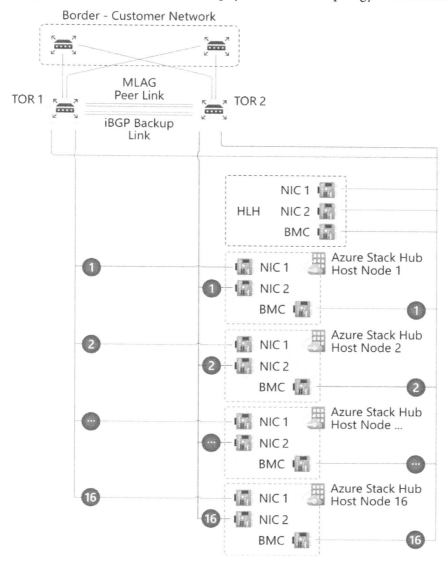

Figure 3.8 – Azure Stack Hub physical network connectivity

Logical networking

Microsoft Azure Stack Hub relies on a series of subnets and each has a different role to play within the platform. The subnets used are as follows:

- **Infrastructure**: Subnet size /24.
- **BMC**: Subnet size /26.
- **Switch infrastructure**: Subnet size /26.
- **Public VIP**: Subnet size must be between /26 and /22.
- **Private**: Subnet size /20.

Let's look at what they are in the next sections.

Infrastructure

The infrastructure network is a dedicated network for internal Azure Stack components to communicate.

BMC

The BMC network is dedicated to connecting all baseboard management controllers. It is used to control physical server power on-off sequences for *OS* installation. It also provides connectivity to the **deployment virtual machine** (**DVM**) during deployment.

The switch management network provides out-of-band access for deployment, management, and troubleshooting.

Public

The public VIP network is assigned to the network controller in the Azure Stack Hub **software defined networking** (**SDN**). The **software load balancer** (**SLB**) assigns /32 networks for tenant workloads and advertises them in the routing table. Azure Stack Hub uses a total of 31 addresses. Eight public IP addresses are used for a small set of Azure Stack Hub services and the rest used by tenant virtual machines. If you plan to use App Service and the SQL resource providers, seven more addresses are used. The remaining 15 IPs are reserved for future Azure services. The network size on this subnet can range from a minimum of /26 to a maximum of /22.

Private

The private network is used for storage, infrastructure containers, private VIP, and other internal functions.

Switch infrastructure

This is for point-to-point IP address routing, dedicated switch management interfaces, and loopback addresses assigned to the switch.

The following diagram shows a summary of the required connections from within the Azure Stack Hub solution:

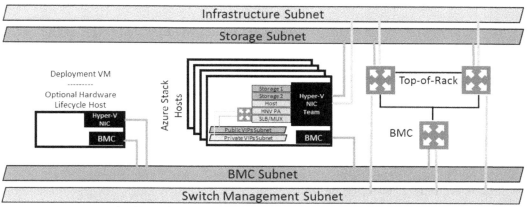

Figure 3.9 – Azure Stack Hub connectivity

Different networks within the stack require differing connectivity, as mentioned earlier in this document.

There is a requirement that several networks will be connected to your local network for access and administrative purposes, to both manage and use the interfaces within Azure, and if using as an internal cloud-based system, access to the server endpoints.

Data center integration touchpoints

There are certain touchpoints for integration into the data center that need to be thought about and used to populate the *PowerShell* UI. These touchpoints will include the following:

- **IPv4**: IP subnets, address spaces
- **Network**: Uplink, BGP, switch vendor, firmware version
- **Firewall**: Publishing rules
- **Identity**: AAD/ADFS
- **SSL**: SSL certificates
- **DNS**: Domain name services
- **NTP**: Time server

In addition to these touchpoints, there are also some other touchpoints that will need to be configured after the deployment is completed. These additional touchpoints include the following:

- **Device authentication**: Radius/TCAS
- **Syslog**: Syslog server
- **ITSM/security**: ITSM integration

These considerations need to be taken into account during the planning stage and worked through with your OEM vendor prior to completing the deployment worksheet, which we will cover later on in this chapter. The planning process will vary depending on the OEM vendor who is deploying Azure Stack Hub, but there are likely to be several planning meetings and assistance for populating the deployment worksheet.

Edge deployment

A network firewall is a recommended way of protecting Microsoft Azure Stack Hub. Azure Stack Hub uses ACLs for the VIPs and physical switches. The following diagram shows how Azure Stack Hub is connected to an existing edge firewall:

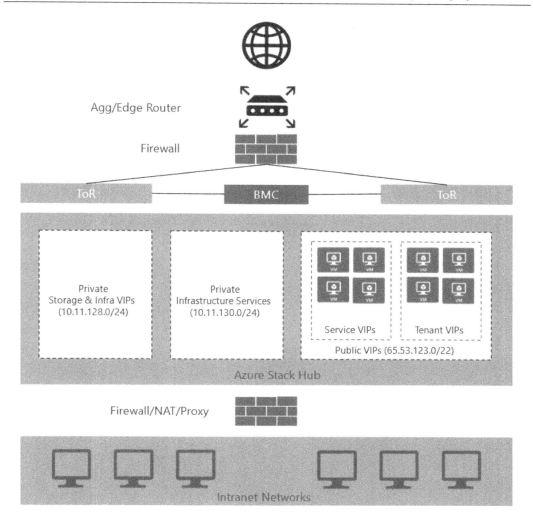

Figure 3.10 – Edge deployment

Enterprise deployment

A traditional deployment in an enterprise would protect Azure Stack Hub by virtue of a **DMZ**, which sits between the internal network and the internet, as shown in the following diagram:

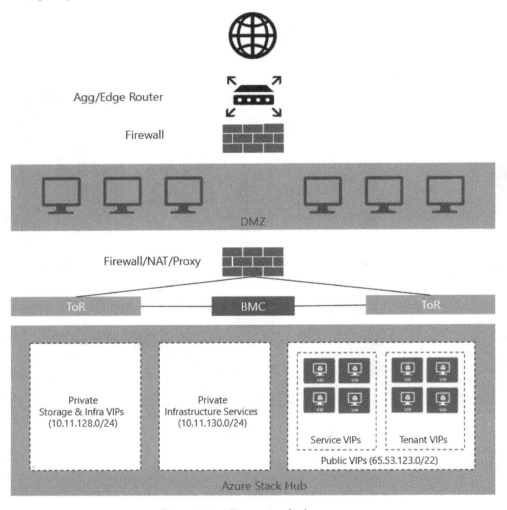

Figure 3.11 – Enterprise deployment

Azure Stack Hub can connect to the **DMZ** through a firewall, via **NAT** or with a proxy.

DNS integration

To be able to access Azure Stack Hub endpoints such as the portal from outside of Azure Stack Hub, integration of the Azure Stack Hub DNS and the DNS servers that host the DNS zones to be used by Azure Stack Hub is needed.

Some information for DNS is populated in the deployment worksheet that we will discuss later in this chapter. This includes the following:

- Region
- External domain name
- Internal domain name
- DNS forwarder
- Naming prefix

The FQDN of the Azure Stack Hub deployment and associated endpoints is a combination of the region parameter and the external domain name parameters. So, for example, for an Azure Stack Hub deployment to `azurecloud.northwinds.com` with a west region, the portal endpoint would be `https://portal.west.azurecloud.northwinds.com`.

To be able to use this example namespace for an Azure Stack Hub deployment, all of the following conditions must be met:

- The `azurecloud.northwinds.com` zone must be registered either with a domain registrar, an internal corporate DNS server, or both, depending on your name resolution requirements.
- The `azurecloud.northwinds.com` child domain must exist under the `northwinds.com` zone.
- The DNS servers that host both the `northwinds.com` and `azurecloud.northwinds.com` zones can be reached from the Azure Stack Hub deployment.

To be able to resolve DNS names of Azure Stack Hub-related endpoints and instances from outside of Azure Stack Hub, you will need to integrate the DNS servers that host the external DNS zone of Azure Stack Hub with the external DNS servers that host the parent zone you wish to use.

There are two types of DNS servers:

- An authoritative DNS server that hosts DNS zones and answers DNS queries for records that exist within those zones only.
- A recursive DNS server does not host any DNS zones and answers all DNS queries by querying the authoritative DNS servers.

An Azure Stack Hub environment includes both authoritative and recursive DNS servers. The recursive DNS servers are used to resolve the names of everything except for the internal private zone and the external public DNS zone used for the Azure Stack Hub deployment.

The authoritative DNS servers are the ones that hold the external DNS zone information, and any user-created zones. You will need to integrate with these servers to enable zone delegation or conditional forwarding to resolve Azure Stack Hub DNS names from outside Azure Stack Hub.

To be able to integrate your Azure Stack Hub deployment with your DNS infrastructure, you will need the following information:

- DNS server FQDNs
- DNS server IP addresses

The FQDNs for the Azure Stack Hub DNS servers all have the following format:

- `<naming prefix>-ns-1.<region>.<externaldomainname>`
- `<naming prefix>-ns02.<region>.<externaldomainname>`

This information is created toward the end of the Azure Stack Hub deployment and is stored in a file named `AzureStackStampInformation.json` on the DVM in the `c:\CloudDeployment\logs` folder.

To be able to resolve DNS names for the endpoints that sit outside of the Azure Stack Hub environment, you must provide reachable DNS servers that Azure Stack Hub can use to forward all DNS requests to for which Azure Stack Hub is not the authority. The values for the DNS forwarder are provided in the deployment worksheet and are required for deployment; otherwise, deployment will fail. These DNS servers can be changed post-deployment using PowerShell with the `set-AzSDnsForwarder` cmdlet.

To enable name resolution with your existing DNS infrastructure, you must configure conditional forwarding. You can add additional forwarders to Azure Stack through the use of PowerShell connected to the **privileged endpoint** (**PEP**). This can be completed using the following cmdlets running as an administrator and connected to the IP address of the PEP using the `CloudAdmin` credentials:

```
$cred = Get-Credential
Enter-PSSession -ComputerName <IP address of ERCS>
-ConfigurationName PrivilegedEnpoint -Credential $cred
Register-CustomDnsServer -CustomDomainName "<domain name>"
-CustomerDnsIPAddresses "ip1,ip2"
```

The easiest and most secure way of integrating Azure Stack Hub with your DNS infrastructure is to do conditional forwarding of the zone from the server that hosts the parent zone. This is the approach that is recommended by Microsoft when you have direct control over the DNS servers that host the parent zone for your Azure Stack Hub deployment.

The final piece to consider with network integration is the integration of firewalls. It is highly recommended that a firewall device is used to help secure the Azure Stack Hub environment. Firewalls can help to defend against things such as **distributed denial-of-service** (**DDoS**) attacks, intrusion detection, and content inspection. Conversely, they can also become throughput bottlenecks for Azure storage services such as blobs, tables, and queues.

Certificate requirements

Azure Stack Hub requires **PKI**, or **public key infrastructure**, certificates with relevant DNS names for public infrastructure endpoints exposed by Azure Stack Hub, and these are required for deployment. There are a series of mandatory certificates that must be in place for deployment as well as optional certificates for other value-added resource providers. In this section, we are going to cover each of these certificates and their requirements in turn.

Microsoft Azure Stack Hub relies on certificates that have been issued by a public trusted **certificate authority** (**CA**) or an **enterprise CA**. Certificates are used for all VIPs that are external-facing and registered in an external **DNS** zone. There is support for wildcard certificates and rotation is conducted via the PEP.

The following certificates are required at deployment time in `PFX` format:

- `Adminportal.<region>.<fqdn>`
- `Portal.<region>.<fqdn>`
- `Adminmanagement.<region>.<fqdn>`
- `Management.<region>.<fqdn>`
- `*.blob.<region>.<fqdn>`
- `*.table.<region>.<fqdn>`
- `*.queue.<region>.<fqdn>`
- `*.adminvault.<region>.<fqdn>`
- `*.vault.<region>.<fqdn>`
- `Adfs.<region>.<fqdn>`
- `Graph.<region>.<fqdn>`
- `*.adminhosting.<region>.<fqdn>`
- `*.hosting.<region>.<fqdn>`

The **ADFS** and **Graph** certificates are only needed when Azure Stack Hub uses ADFS for identity. Microsoft provides a guide for the certificates that are required for Azure Stack Hub deployment and registration at the following URL: `https://docs.microsoft.com/en-us/azure-stack/operator/azure-stack-pki-certs?view=azs-2102`.

The Azure Stack Hub Readiness Checker tool can be used to create **certificate signing requests** (**CSRs**) that are suitable for an Azure Stack Hub deployment.

The Azure Stack Hub Readiness Checker tool can be installed from within a PowerShell window running as an administrator using the following command:

```
install-Module Microsoft.AzureStack.ReadinessChecker
```

The subject, output directory, identity, region name, and external FQDN for the Azure Stack Hub solution can be set as variables that will be passed to the signing request as shown here:

```
$subject = "C=US,ST=Washington,L=Redmond,O=Microsoft,OU=Azure Stack Hub"
$outputDirectory = "$ENV:USERPROFILE\Documents\AzureStackCSR"
$IdentitySystem = "AAD"
```

```
$regionName = "east"
$externalFQDN = "azurestack.contose.com"
New-AzsHubDeploymentCertificateSigningRequest -RegionName
$regionName -FQDN $externalFQDN -subject $subject
-OutputRequestPath $outputDirectory -IdentitySystem
$IdentitySystem
```

This command will generate a .REQ file, which should then be submitted to your CA.

Once the certificates have been returned from the CA, they should be copied to a single directory on the same machine from which the CSR was submitted. The Azure Stack Hub Readiness Checker tool can then be used to package the certificates ready for deployment using the following cmdlets:

```
$Path = "$env:USERPROFILE\Documents\AzureStack"
$pfxPassword = Read-Host -AsSecureString -Prompt "PFX Password"
$ExportPath = "$env:USERPROFILE\Documents\AzureStack"
ConvertTo-AzsPFX -Path $Path -pfxPassword $pfxPassword
-ExportPath $ExportPath
```

This step will export and validate the certificates. Once validation is complete, they can be presented for deployment.

The Azure Stack Hub Readiness Checker tool can also be used to validate that your Azure subscription is ready to use with Azure Stack Hub before the deployment commences. This validates that the Azure subscription is of the correct type and the account to be used to register the subscription with Azure can sign in to Azure and owns the subscription.

The following cmdlets can be run from within a PowerShell window running as an administrator:

```
$subscriptionID = "your subscription id"
Connect-AzAccount -subscription $subscriptionID
Invoke-AzsRegistrationValidation -RegistrationSubscriptionID
$subscriptionID
```

This should return an okay status if all requirements for registration have been met.

Crucial decisions

There are some decisions that must be thought about before deployment as if they are changed, then this may mean either a complete re-deployment or others that will be difficult to implement without a rebuild.

Changes that would require a re-deployment of Azure Stack Hub include the following:

- Switching between AAD and ADFS for identity management
- Switching to a different AAD tenant name for the primary/default tenant
- Changing the name of the Azure Stack Hub region or any part of the FQDN
- Changing from static routing to BGP or vice versa
- Changing the internal domain name used for the Active Directory within the Azure Stack Hub
- Changing the initial public IP addresses
- If the incorrect hardware was chosen initially
- If too many nodes were chosen initially

Some other changes could be made without a complete re-deployment but may be painful to implement either from a technical standpoint or a business process point of view. These include the following:

- If you have plans and offers that have been accepted by tenants that you decide are incorrect
- Changing ADFS to a new external Active Directory
- If you underestimate the virtual machine workload and need to order additional nodes (not complicated but will take time to order and implement)
- Changing from a capacity billing model to a consumption billing model (you would still need to pay for capacity for the current period).
- Changing registrations to Azure subscriptions and Enterprise Agreement (EA) isn't painful but historical billing information is not in one consolidated EA.

As you can see, it is important to ensure that these decisions are made correctly at the time of the first deployment to save problems further down the line that may incur additional costs and platform downtime.

When you are looking at implementing Microsoft Azure Stack Hub within your own data center, you will work with your chosen OEM vendor who is supplying your solution. As part of working with your OEM vendor, they will collect a series of information from you. This includes information for names, IP ranges, and BGP routing, which would be required for a fully automated configuration of Microsoft Azure Stack Hub. This is a PowerShell-based UI that collects the information and produces a set of configuration files. The PowerShell UI is downloadable from the web or it may be provided to you by the OEM vendor that is supplying the Azure Stack Hub solution.

The easiest way to download and install the deployment worksheet is to do it directly from within PowerShell using the following from a PowerShell **integrated scripting environment** (**ISE**) running as an administrator:

```
install-Module -Name Azs.Deployment.Worksheet
```

This can then be run from within the same PowerShell window with the following:

```
start-DeploymentWorksheet
```

The following screenshots show an example of the forms from the PowerShell UI:

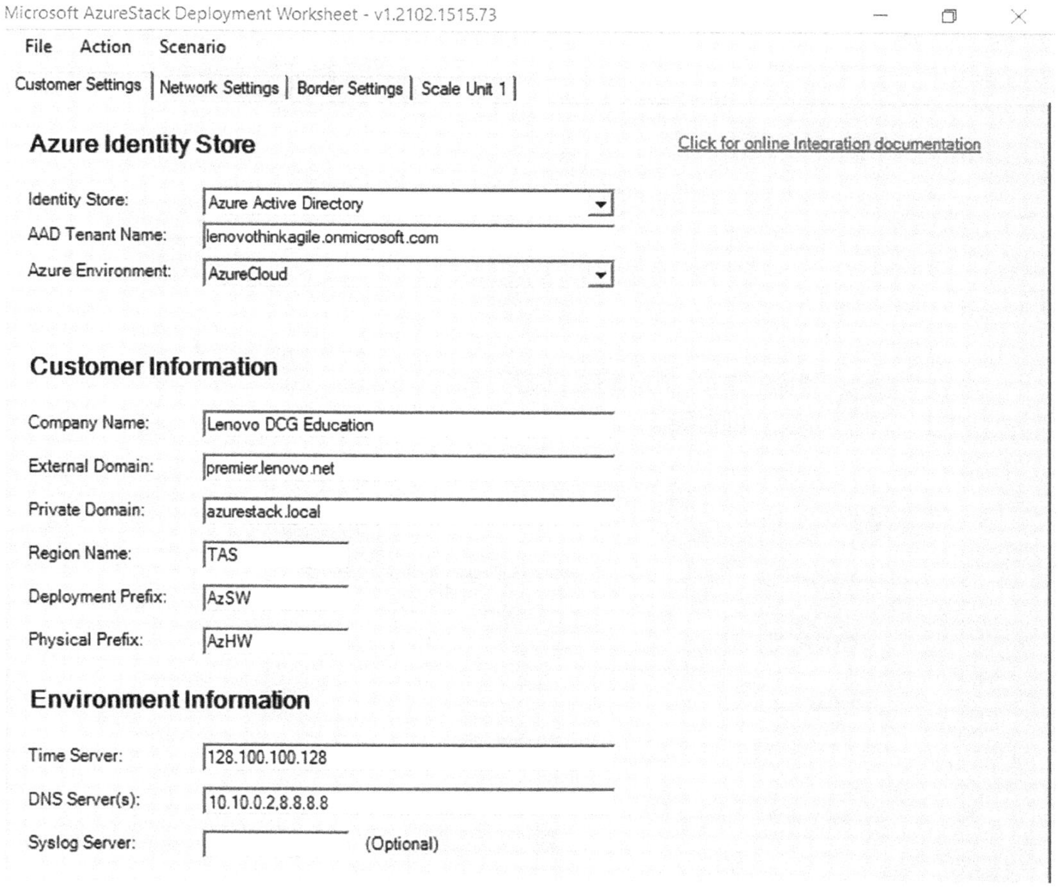

Figure 3.12 – PowerShell UI customer and identity form

This first preceding screenshot shows the initial **Customer Settings** tab where we specify the identity store and domain information.

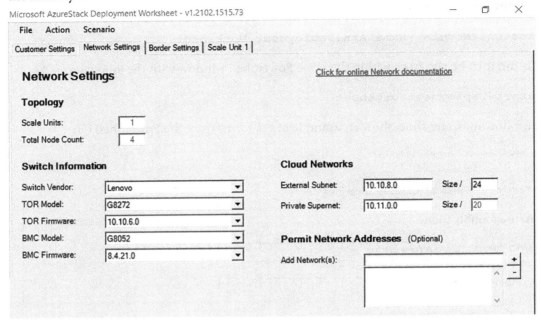

Figure 3.13 – PowerShell UI Network Settings form

Figure 3.13 shows the form that details the network switch information and the topology.

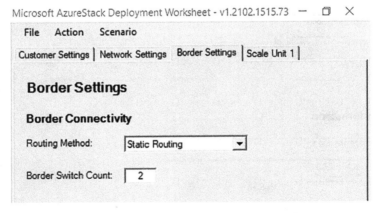

Figure 3.14 – Border Settings PowerShell UI form

Figure 3.14 shows the form that details the border connectivity, either static or BGP.

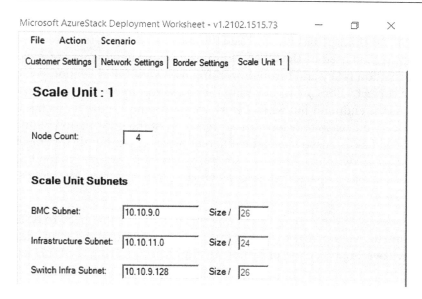

Figure 3.15 – Scale Unit settings PowerShell UI form

Once the PowerShell forms are populated, the customer will then select **Action**, and then click on **Generate** on the drop-down menu to create the settings file data.

Performing this action from within the PowerShell UI will generate the settings file as per the next screenshot:

Figure 3.16 – Settings file generated from the PowerShell UI

The settings file that is generated from the PowerShell UI is then shared with the OEM vendor who will use this settings file to create the automated deployment script. This generated script is then used by the OEM vendor during the deployment of Azure Stack Hub. This information is key and through working with your OEM vendor, you can ensure that the correct values are used within the worksheet to ease the deployment. The worksheet is the endpoint but when I work with customers, I use it as a talking point from the very outset so they have an appreciation of the information required for the deployment and as a consequence, the different pieces of the puzzle that need to be considered, such as networking. There are several touchpoints for integration that need to be thought about and prepared before deployment can be initiated. We will cover these touchpoints now.

Now that we understand the prerequisites and crucial decisions that need to be taken prior to deployment, we can start to look at the process involved in the actual deployment. We will start by covering two types of deployment models, featuring the **HLH** and the **DVP**.

Getting to know HLH and DVP deployment

In this section, we will take a look at the deployment setup required before the actual deployment of the Azure Stack Hub platform and two key components of the deployment process. These are the HLH and DVP.

This is important as Azure Stack Hub deployments will not be performed by end customers, unlike most traditional on-premises products or solutions from Microsoft. The acquisition and deployment of Azure Stack Hub require a relationship between Microsoft and the selected OEM partner.

The Azure Stack Hub solution is provided as an integrated system from the OEM vendor. The OEM vendor is then responsible for the following deployment activities:

- Hardware deployment
- OEM and Microsoft support onboarding
- Azure Stack Hub platform deployment
- Basic integration services (such as networking)

For an Azure Stack Hub multi-node deployment, there are two key components required for the deployment process. These are both shown in the following diagram:

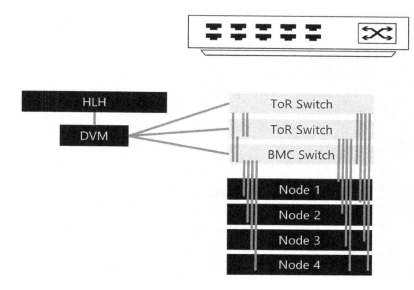

Figure 3.17– Multi-node Azure Stack Hub deployment platform

You can see in *Figure 3.17* the HLH and associated DVMs that are used during the deployment process.

Hardware life cycle host (HLH)

The HLH is a separate physical machine used for deployment and shipped with each scale unit. This is used for the initial Azure Stack Hub deployment and can also be used for third-party applications from the hardware vendor. It is worth noting that not all hardware vendors will have an HLH as part of their offering. For example, *Cisco* does not ship with an HLH. From a deployment standpoint, typically the hardware vendor will connect to the HLH on-site to kick off the deployment process of Azure Stack Hub. If the customer or vendor so desires, then remote access can be granted to the HLH and the deployment process can then be kicked off remotely.

Deployment virtual machine (DVM)

The DVM is built as a *Hyper-V* virtual machine that is run on the HLH. The actual deployment of Azure Stack Hub is started from this virtual machine. During the setup process of Azure Stack Hub, this virtual machine will temporarily act as AD, DHCP, WDS, and others. Once Azure Stack Hub has successfully been deployed, the DVM is deleted.

The DVM is created by the OEM vendor on the HLH by virtue of a PowerShell script, which you can see an example of in the following screenshot:

```
.\InitializeAzureStackDeployment.ps1 `
    -ComputerName <dvmname> `
    -LocalAdministratorPassword <dvmpass> `
    -IPAddress <dvmip> `
    -DVMHostMACAddress <mac> `
    -NetMask <subnetmask> `
    -DefaultGateway <dvmgateway> `
    -VlanId <vlanid> `
    -OemIsoPath <oemiso> `
    -Verbose
```

Figure 3.18 – Initialization script to create a DVM

The script allows the customer to provide details for the parameters as per their infrastructure and setup.

After the creation of the DVM on the HLH, the actual deployment of Azure Stack Hub can commence. This is what we are now about to cover in the next section.

Deploying Azure Stack Hub

The installation or deployment of Azure Stack Hub can be started once all of the prerequisites have been completed, including the creation of the configuration files from the PowerShell UI that we looked at previously. All this information will have been shared with the OEM vendor, who would then have shipped the hardware required for Azure Stack Hub. The onsite engineer from the OEM vendor will perform the following tasks during the deployment:

- Rack the solution.
- Integrate to the customer network.
- Prepare the HLH.
- Update all the BMC and firmware to all hosts.
- Kick off the deployment.
- Install the latest updates.
- Test the Azure Stack Hub post-deployment.
- Register Azure Stack Hub.
- Hand off the solution to the customer.

Microsoft is available to the engineer for support to help with any issues that may arise during the deployment. Certainly, in my experience, when there are issues during deployment, generally it is down to a misconfiguration of networking.

A typical engagement with an OEM vendor for the deployment of an Azure Stack Hub solution will include 2 days of remote assistance with preparation, including network changes, certificates, deployment worksheet population, and readiness checking. This is then followed by 4 days onsite performing the deployment and handing over, depending on the number of nodes to be deployed.

The beginning of the Azure Stack Hub deployment is started from the DVM, which is running on the HLH. The driver package to be included is specified when the DVM is created. There are different parameters required depending on whether AAD or ADFS is selected for identity management. The rest of the configuration parameters are either passed directly to the process, or more commonly, passed as a JSON file that has been generated from the deployment PowerShell UI.

As with the creation of the DVM, the installation of Azure Stack Hub is controlled by a PowerShell script – an example of which is as follows:

```
InstallAzureStack.ps1 `
    -InfraAzureEnvironment "AzureCloud" -CompanyName "<..>" `
    -InfraAzureDirectoryTenantName "<..>.onmicrosoft.com" `
    -InfraAzureDirectoryTenantAdminCredential <..> `
    -DomainFQDN <..> -DomainAdminCredential <..> `
    -BareMetalCredential <..> -NamingPrefix <..> `
    -TimeZone <..> -TimeServer <..> -EnvironmentDNS <..> `
    -TORSwitchBGPASN <..> -SoftwareBGPASN <..> `
    -TORSwitchBGPPeerIP <..> -InfrastructureNetwork @{Subnet=<..>} `
    -StorageNetwork @{Subnet=<..>; vlanId=<..>} `
    -InfrastructureExtendedNetwork @{Subnet=<..>} `
    -ExternalNetwork @{Subnet=<..>} -RegionName "<..>" `
    -PhysicalNodes @( @{ Name=".."; BMCIPAddress="<..>"; MACAddress="<..>"},
                      @{ Name=".."; BMCIPAddress="<..>"; MACAddress="<..>" },
                      @{ Name=".."; BMCIPAddress="<..>"; MACAddress="<..>" },
                      @{ Name=".."; BMCIPAddress="<..>"; MACAddress="<..>" } )
```

Figure 3.19 – Install Azure Stack Hub PowerShell script

It is worth noting that the installation will be performed by your OEM vendor and it is detailed here for clarity.

Now that we have had a look at the parameters and the PowerShell script, let's visit the definition of physical nodes.

Azure Stack Hub nodes and names

Nodes are defined as physical nodes and each node represents a server within a scale unit. Each node will have its own name, IP address, and MAC address. The name is autogenerated using a defined prefix but can be overwritten in the PowerShell script. An example of defining the nodes within the PowerShell script is shown in the following screenshot:

```
$node1 = @{ Name=„azs-node1"; BmcIPAddress="10.10.34.5"; MacAddress="AB-CD-EF-12-34-56"}
$node2 = @{ Name=„azs-node2"; BmcIPAddress="10.10.34.6"; MacAddress=".."}
$node3 = @{ Name=„azs-node3"; BmcIPAddress="10.10.34.7"; MacAddress=".."}
$node4 = @{ Name=„azs-node4"; BmcIPAddress="10.10.34.8"; MacAddress=".."}
...

$physicalNodes = @( $node1, $node2, $node3, $node4, ..)
```

Figure 3.20 – Node definition in PowerShell

The prefix (`azs` in the preceding example) will come from the configuration files that were generated by the PowerShell UI that we went through at the beginning of this chapter.

Rerun

In normal circumstances, the PowerShell script to deploy Microsoft Azure Stack Hub only needs to be run once. Sometimes there are intermittent environmental issues that cause the deployment to fail and to assist with these circumstances, there is an additional parameter that can be used, called `-ReRun`.

The `-ReRun` parameter will trigger a re-run of the deployment script. With this parameter included, the deployment script will skip steps that have been successfully completed on the previous run of the deployment script and will start from the step where it previously failed. It is vitally important that no credentials or parameters are changed. If credentials need to be changed, then it is best practice to reset the environment and then restart the deployment from the beginning again.

We now understand the deployment process and the important switches for the PowerShell script responsible for the deployment. In the next section, we are going to cover some post-deployment tasks and provide some quick troubleshooting tips should something go wrong during deployment.

Post-deployment and troubleshooting

This section goes through some of the changes that need to be performed after a successful deployment of Azure Stack Hub, which includes validating the Azure Stack Hub solution and registering with Azure.

Post-deployment tasks

Once the Azure Stack Hub deployment is complete, then there are a few things that need to be completed before we can be sure that the Azure Stack Hub environment is operational and ready for hand over to the customer.

The following tasks can be run post-deployment:

- Test Azure Stack.
- Register Azure Stack.
- Populate Marketplace.
- Install additional resource providers.
- Configure quotas/plans/services.
- Configure RBAC.

We will be covering the first two in this chapter and we will be working through the remainder as we continue through the rest of the book.

The first task is the running of the `test-azurestack` PowerShell cmdlet, which will help to detect failures that don't result in cloud outages, such as a single failed disk or a single physical node failure. A sample of the output from the `test-azurestack` PowerShell command is shown in the following screenshot:

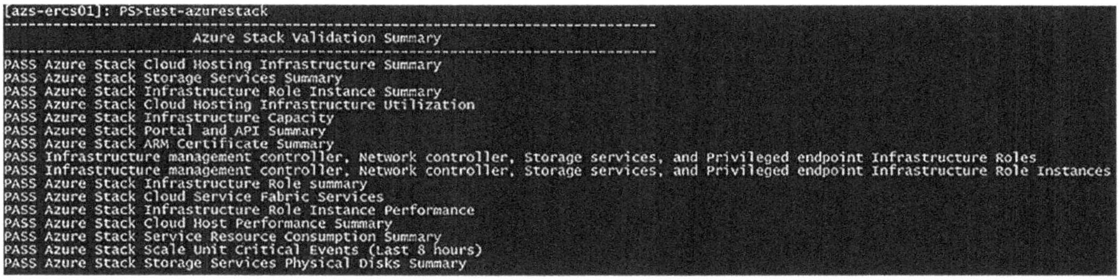

```
[azs-ercs01]: PS>test-azurestack
--------------------------------------------------------------------------------
                       Azure Stack Validation Summary
--------------------------------------------------------------------------------
PASS Azure Stack Cloud Hosting Infrastructure Summary
PASS Azure Stack Storage Services Summary
PASS Azure Stack Infrastructure Role Instance Summary
PASS Azure Stack Cloud Hosting Infrastructure Utilization
PASS Azure Stack Infrastructure Capacity
PASS Azure Stack Portal and API Summary
PASS Azure Stack ARM Certificate Summary
PASS Infrastructure management controller, Network controller, Storage services, and Privileged endpoint Infrastructure Roles
PASS Infrastructure management controller, Network controller, Storage services, and Privileged endpoint Infrastructure Role Instances
PASS Azure Stack Infrastructure Role summary
PASS Azure Stack Cloud Service Fabric Services
PASS Azure Stack Infrastructure Role Instance Performance
PASS Azure Stack Cloud Host Performance Summary
PASS Azure Stack Service Resource Consumption Summary
PASS Azure Stack Scale Unit Critical Events (Last 8 hours)
PASS Azure Stack Storage Services Physical Disks Summary
```

Figure 3.21 – test-azurestack PowerShell output

Figure 3.21 shows a healthy Azure Stack Hub deployment with no failures detected.

If the `test-azurestack` cmdlet returns an error, then it is possible to re-run the cmdlet with the `-Debug` switch to see greater detail in the output.

The next step that needs to happen once the `test-azurestack` PowerShell has been run is to register the Azure Stack Hub instance. This needs to be done regardless of whether this is a connected or disconnected scenario. The process will be slightly different between the connected and disconnected scenarios.

There are some prerequisites that need to be obtained prior to registering the Azure Stack Hub instance. These include details for the Azure subscription, the billing model (capacity or consumption), and the deployment ID that comes from Azure Stack Hub, which has just been deployed. As with the deployment and testing of Azure Stack Hub, the registration process is also performed in PowerShell.

The PowerShell Azure Stack Hub PEP is a remote PowerShell console that comes pre-configured and provides you with just enough capabilities to perform the required tasks. This endpoint exposes only a restricted set of cmdlets through the use of PowerShell **Just Enough Administration (JEA)**. The PEP is used during registration and may also be used with Microsoft support to perform certain tasks while troubleshooting issues.

Connected registration

To be able to register Azure Stack Hub successfully, the PowerShell language mode has to be set to `FullLanguageMode`.

Run the `Connect-AzAccount -EnvironmentName "<environment name>"` cmdlet from a PowerShell ISE window, which is run as an administrator. The value for the environment name parameter will be one of `AzureCloud` or `AzureUSGovernment`.

Next, run `Register-AzResourceProvider -ProviderNamespace Microsoft.AzureStack` from the same PowerShell window.

The registration process relies on the `RegisterWithAzure` PowerShell module, which is part of the `AzureStack-Tools` GitHub repository. The repository can be downloaded using the following script:

```
Cd \
[Net.ServicePointManager]::SecurityProtocol =
[NetSecurityProtocolType]::Tls12
invoke-webrequest 'https://github.com/Azure/AzureStack-Tools/
archive/az.zip' -OutFile az.zip
expand-archive az.zip -DestinationPath . -Force
cd AzureStackTools-az
```

This will then allow us to import the `RegisterWithAzure` PowerShell module using the following:

```
Import-Module .\RegisterWithAzure.psm1
```

Then, to perform the registration for the pay-as-you-use billing model, execute the following cmdlet from the same PowerShell window:

```
$CloudAdminCred = Get-Credential -UserName <Privileged endpoint
credentials> -Message "Enter the cloud domain credentials to
access the privileged endpoint."
$RegistrationName = "<unique-registration-name>"
Set-AzsRegistration '
    -PrivilegedEndpointCredential $CloudAdminCred '
    -PrivilegedEndpoint <PrivilegedEndPoint computer name> '
    -BillingModel PayAsYouUse '
    -RegistrationName $RegistrationName
```

The same command is used for the capacity billing model, although the syntax is slightly different, as shown here:

```
$CloudAdminCred = Get-Credential -UserName <Privileged endpoint
credentials> -Message "Enter the cloud domain credentials to
access the privileged endpoint."
$RegistrationName = "<unique-registration-name>"
Set-AzsRegistration '
    -PrivilegedEndpointCredential $CloudAdminCred '
    -PrivilegedEndpoint <PrivilegedEndPoint computer name> '
    -AgreementNumber <EA agreement number> '
    -BillingModel Capacity '
    -RegistrationName $RegistrationName
```

This registration should take around 15 minutes to complete and the message `Your environment is now registered and activated using the provided parameters` should be displayed.

Disconnected registration

The registration process for a disconnected Azure Stack Hub is slightly different from that of the connected mode due to the lack of internet connectivity.

As with the connected registration process, the registration is reliant on `AzureStack-Tools` from GitHub and the `RegisterWithAzure.psml` module is imported in a PowerShell ISE window running as an administrator as detailed previously.

With the disconnected registration process, a registration token is required, which is achieved using the following cmdlet:

```
$FilePathForRegistrationToken = "$env:SystemDrive\
RegistrationToken.txt"

$RegistrationToken = Get-AzsRegistrationToken
-PrivilegedEndpointCredential $YourCloudAdminCredential
-UsageReportingEnabled:$False -PrivilegedEndpoint
$YourPrivilegedEndpoint -BillingModel Capacity
-AgreementNumber '<EA agreement number>' -TokenOutputFilePath
$FilePathForRegistrationToken
```

This registration token should be saved and copied across to an internet-connected machine for use in the registration to Azure.

From this internet-connected machine, again use a PowerShell ISE window running as an administrator. Download the GitHub repository and import the `RegisterWithAzure` PowerShell module.

The following PowerShell cmdlets are then run within this window:

```
$RegistrationToken = "<Your Registration Token>"
$RegistrationName = "<unique-registration-name>"
Register-AzsEnvironment -RegistrationToken $RegistrationToken
-RegistrationName $RegistrationName
```

This registers Azure Stack Hub but as the instance is disconnected, an activation key is required, which is retrieved from Azure using the following PowerShell cmdlets:

```
$RegistrationResourceName = "<unique-registration-name>"
$KeyOutputFilePath = "$env:SystemDrive\ActivationKey.txt"
$ActivationKey = Get-AzsActivationKey -RegistrationName
$RegistrationResourceName -KeyOutputFilePath $KeyOutputFilePath
```

The activation key created with this command is then copied back to Azure Stack Hub and from a PowerShell ISE running as an administrator, run the following PowerShell cmdlet to complete the registration:

```
$ActivationKey = "<activation key>"
New-AzsActivationResource -PrivilegedEndpointCredential
$YourCloudAdminCredential -PrivilegedEndpoint
$YourPrivilegedEndpoint -ActivationKey $ActivationKey
```

For either of these registration processes, you can verify that it was successful by using the **Region management** tile within the administrator portal. Selecting properties from the **Region management** tile will show the subscription status, which if the registration process was successful will show **Registered**. The other potential values for this are **Not registered** or **Expired**. A successfully registered Azure Stack Hub will be shown, as per the following screenshot, in the administrator portal:

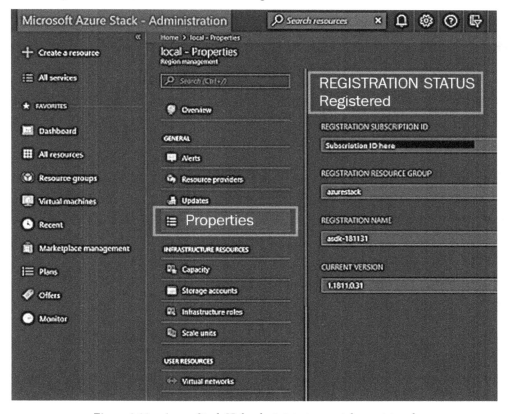

Figure 3.22 – Azure Stack Hub administrator portal – registered

There are some events that will trigger a re-registration or update of registration of the Azure Stack Hub instance. These include the yearly renewal of a yearly capacity-based subscription. A change in the billing model between capacity and consumption will also require a re-registration, as will the adding of additional nodes in the capacity-based billing model.

Installation of the additional resource providers is not included by default as part of the core Azure Stack Hub deployment. These additional resource providers include SQL, MySQL, Event Hubs, and IoT Hubs, among others. If any of these are required, then they would need to be downloaded and installed separately. We will touch on these resource providers later in the book but it is enough for now to know that these are not part of the core Azure Stack Hub deployment.

There are several tools available on GitHub, such as adding virtual machine items to the gallery. That is beyond the scope of this book, but the tools can be found here: `https://github.com/Azure/AzureStack-Tools`.

As part of the Azure Stack Hub deployment, the OEM vendor will be responsible for agentless hardware device monitoring. Each hardware vendor will have specific solutions for this. As an example, Lenovo provides XClarity, which features controllers in the hardware and an administration portal that is deployed to the HLH. This provides call home functionality for failures such as disk drives, which will automatically order a new disk. Other vendors will have similar solutions deployed to the HLH during their Azure Stack Hub deployments.

Integration with monitoring solutions

There are several touchpoints that can be integrated into existing monitoring solutions. Some of these are going to be hardware vendor-dependent but Microsoft Azure Stack Hub has some already built-in. The key touchpoints are as follows:

- Azure Stack Hub software API:

  a) Included as part of **System Center Operations Manager (SCOM)** – management pack

  b) The *Nagios* plugin

  c) API examples available on *GitHub*

- Physical server – **BMC**:

  a) SCOM – hardware vendor management pack

  b) Hardware vendor Nagios plugins (Lenovo uses XClarity, for example)

  c) OEM-supported monitoring solutions

- Network devices – **SNMP**:

  a) SCOM – network devices discovery

  b) Nagios switch plugin

  c) OEM-supported network monitoring solutions

- Tenant subscription health monitoring:

  a) **SCOM**: Azure management pack

  b) **Operations Manager Suite (OMS)**

Tools and tips

The Azure Stack Hub Tools repository provides several tools that can be used alongside the administration portal to automate some repetitive tasks. This includes the following modules:

- Syndication – offline marketplace syndication

- Policy

- Data center integration

Tools available from this repository include the TemplateValidator tool, which can be used to safeguard against hardcoded locations and hardcoded storage endpoints within your ARM templates. We will be covering ARM templates later in the book. There is also the Azure Stack Policy tool, which safeguards against unsupported resource types being deployed to Azure Stack Hub from Azure.

Troubleshooting

As mentioned earlier in this chapter, it is possible to re-run the Azure Stack Hub deployment PowerShell script with the -Rerun switch. But what happens if there are issues post-deployment once the OEM vendor technician has left? Most OEM vendors will offer some sort of support for such eventualities. As part of this support, you are likely to have to provide logs from the Azure Stack Hub environment to aid with the **troubleshooting**. Fortunately, Microsoft makes this straightforward with a PowerShell cmdlet that can collect all of the relevant Azure Stack Hub logs and make them available to the OEM. The cmdlet is Get-AzureStackLog, which would need to be run against a PEP. If this cmdlet is run shortly after deployment is complete, then it will provide details of the deployment and any issues that were present at the point of deployment. We will dive deeper into troubleshooting and support in a later chapter in the book but if there are any issues during deployment, then the first place to start is the *log*, which is viewable from the Azure Stack Hub admin portal.

Summary

In this chapter, we have focused on the deployment process that is run by the OEM vendor to deploy Azure Stack Hub within an organization's data center. We have talked about the prerequisites and the crucial decisions that need to be taken prior to the deployment process taking place. We have looked through the PowerShell UI, which is used to generate the configuration files used by the OEM vendor during the deployment process.

We have looked at the differences between the connected and disconnected scenarios. Importantly, we now understand what decisions cannot be changed post-deployment. We have talked about the different identity providers that can be configured during deployment. We should now be able to describe both the HLH and the DVM. We should be able to describe the part that each of these plays during the deployment process. We know that the deployment process is performed by the OEM vendor and that it is performed using PowerShell.

We understand the PowerShell script and the different parameters that can be passed. We should be able to describe the different options for passing the parameters to the PowerShell script. We should also be able to describe the switches required to support ADFS and also the switch to perform a rerun of the deployment process if required. We have talked about the registration process to activate the Azure Stack Hub instance.

We have also talked about the post-deployment tasks that need to be undertaken once Azure Stack Hub is operational. We have also looked at integrating Azure Stack Hub into existing monitoring systems such as Microsoft SCOM. In the final part of the chapter, we looked at some tools and tips for troubleshooting both during and after deployment. I would expect some of this information to appear in the questions for those who take the *AZ-600* exam.

This completes the chapter on the deployment of Azure Stack Hub in the book and we will be moving on to the next chapter with a deeper dive into *Azure Stack Hub identity*.

Section 2: Identity and Security

The second section of this book concentrates on identity options and the integration of Microsoft Azure Stack Hub into a modern data center. By the end of this section, you should be more than familiar with all aspects of identity management and security in a hybrid cloud scenario.

The following chapters will be covered under this section:

- *Chapter 4, Exploring Azure Stack Hub Identity*
- *Chapter 5, Securing Your Azure Stack Hub Instance*
- *Chapter 6, Considering DevOps in Azure Stack Hub*

4
Exploring Azure Stack Hub Identity

This chapter offers an in-depth look at the options that can be utilized for identity management within **Microsoft Azure Stack Hub**. It covers both the connected and disconnected scenarios for Microsoft Azure Stack Hub. It details the different ways of communicating with **Active Directory** (**AD**) both on-premises and in Microsoft Azure. We will touch on both **Azure Active Directory** (**AAD**) and **Active Directory Federation Services** (**ADFS**). This chapter introduces some key terminology that we will see more of as we progress through the rest of this book. Azure Stack Hub identity is a key component that will feature in the **AZ-600** exam and will be touched on again in later chapters.

In this chapter, we will be covering the following main topics:

- Introducing the terminology and background
- Understanding the Azure identity model
- Reviewing Azure Stack Hub identity fundamentals
- Overviewing directory-based authentication in depth
- Configuring ADFS and Graph integration
- Deciding between AAD and ADFS

Technical requirements

You can view this chapter's code in action here: `https://bit.ly/2Wg2Wjq`

Introducing the terminology and background

To begin with, I want to cover some terminologies and some background that are necessary for identity management with Microsoft Azure Stack Hub. You might have already seen some of these terms used in the first three chapters of the book, and you will continue to see them throughout the rest of this book.

Directory tenants

Directory tenants are organizations and are the main container within Azure Stack Hub for information on users, applications, and service principals.

Users

Users or **identities** are the people within the organization who are going to be using the applications and services from Azure Stack Hub. They are standard accounts that authenticate individuals with a username and password. The way users within Azure Stack Hub are created and managed depends on the identity provider that has been chosen, either AAD or ADFS. However, we will discuss this in more detail later in the chapter.

Applications

Applications are services that are offered to an organization's users to interact with the application from Azure Stack Hub, such as virtual machines. They are registered to the directory tenant to make them visible to the users.

Service principals

A **service principal** is an application instance of a global application object defined in a single tenant or directory. It is a concrete instance created from the application object and inherits certain properties from the application object. Each tenant, where the application is used and references the global application, will require its own service principal. The service principal defines what the application can do in the relevant tenant, what resources the application can access, and which users are able to use the application.

Services

Services, such as applications, need to be registered with the directory tenant in Azure Stack Hub, which enables the service to authenticate to the identity provider.

Consent

Consent to applications consists of two basic kinds. One is **user-level consent**, which is applicable only to the current user, while **admin-level consent** is applicable to the entire directory. Azure Stack Hub is configured in such a way that it requires an *administrator* to give their consent when connecting to a directory. This is because Azure Stack Hub requires permissions to able to read directory data.

We have covered some of the basic terminologies, so let's turn our attention to the protocols within Azure Stack Hub.

Protocols

Azure Stack Hub allows the use of different protocols to authenticate and authorize users. Here are some of the common ones.

OAuth 2.0

The **OAuth** 2.0 protocol is used to authorize access to web applications and Web APIs. When an application is registered to Azure Stack Hub with OAuth, the application is given an application ID that enables the *application* to receive requests based on this application ID.

OpenID Connect

OpenID Connect is a simple identity layer that is built on top of the OAuth 2.0 protocol. It describes the mechanics used to obtain and use **access tokens** to allow access to protected resources. It adds authentication as an extension of the OAuth 2.0 authorization process. It utilizes the **JavaScript Object Notation** (**JSON**) format. Notably, Azure Stack Hub only uses OpenID Connect internally.

SAML 2.0

The **security assertion markup language** (**SAML**) 2.0 protocol is used when a customer uses ADFS for authentication between the customer ADFS and Azure Stack Hub internal ADFS.

From protocols, we can now move on to token types, which are passed by the protocols for authentication and authorization.

Token types

Azure Stack Hub makes use of tokens at many layers through the protocols we just covered to protect resources. Here are the most common token types used in Azure Stack Hub.

Access tokens

Access tokens are used to access protected resources within Azure Stack Hub. They are used for a specific combination of user, client, and resource. They cannot be revoked and are short-lived tokens.

Refresh tokens

Whenever a resource is accessed, the client receives refresh and access tokens. The refresh token is used to access a new access token when the old one expires. Unlike the access token, the refresh token can be revoked and will usually be valid for a longer period.

ID tokens

An **ID token** contains profile information about a specific user. It remains valid until it expires and is typically represented as a **JSON Web Token**, which contains claims.

Session tokens

A session token is a cookie that is used for **Keep-Me-Signed-In (KMSI)**. These can be revoked; additionally, they can either be persistent or non-persistent.

Now that we have a better understanding of the terminologies of the identity provider and management process within Azure Stack Hub, it is time to move on to look at the Azure identity model.

Understanding the Azure identity model

Microsoft Azure Stack Hub uses identity at different layers. This includes portals and tools that sit on top of **Azure Resource Manager (ARM)**. Moving further down the stack, we also use identity for resource providers. We also use identity for business logic and infrastructure, as shown in the following diagram:

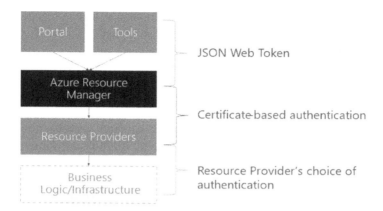

Figure 4.1 – Azure identity layers

As you can see from this diagram, the tokens and access tokens used across the layers are different depending on which layer of Azure Stack Hub we are talking to.

For applications and users, the architecture of Azure Stack Hub is broken down into four layers, as shown in the preceding diagram. Interactions between these layers can use different types of authentication.

For tools and clients such as the administrator portal, the authentication used to communicate to the ARM is via a JSON Web Token. ARM validates the JSON Web Token and looks at the claims issued in the JSON Web Token to check the level of authorization for that user.

The ARM communicates with the resource provider to transfer communications from the user. Calls to the resource provider are secured through the use of **certificates**. Transfers from ARM can be direct imperative calls or declarative calls and are controlled via ARM templates.

Resource providers communicate with the business logic and infrastructure layer using an authentication provider of their choice. Typically, the default resource provider used is **Windows Authentication**, as this is the resource provider that ships with Azure Stack Hub out of the box.

Azure Stack Hub uses **Role-Based Access Control** (**RBAC**), and this is consistent with the implementation of the Azure public cloud. You can manage access to resources by assigning the appropriate RBAC role to users, groups, and applications.

After the deployment of Azure Stack Hub, the AAD global administration permission is not required for Azure Stack Hub. There are instances, however, where this global administration permission may be required to perform some operations within Azure Stack Hub, such as a new feature requiring permission to be granted. It is possible, in this scenario, to temporarily reinstate the account's global admin permission or use a separate global admin account that is an owner of the default provider subscription.

You should now have an understanding of the Azure Stack Hub identity model. So, we can move on and take a look at some Azure Stack Hub identity fundamentals. Please join me in the next section.

Reviewing Azure Stack Hub identity fundamentals

We have already touched on one of the Azure Stack Hub identity fundamentals in *Chapter 1, What Is Azure Stack Hub?*, and *Chapter 2, Azure Stack Architecture*, that is, your choice of identity provider. There are two possible options to select from for the identity resource provider. These are AAD or ADFS. This choice needs to be made prior to deployment, and you cannot change an identity provider post-deployment without a complete redeployment of the Azure Stack Hub solution.

The decision around your identity provider is vitally important as the identity provider you choose might limit your options, particularly around the support of multi-tenancy.

Let's take a look at the differences for each provider:

Capability or scenario	Azure Active Directory (AAD)	Active Directory Federation Services (ADFS)
Connected to the internet?	Yes	Optional
Support for multi-tenancy?	Yes	No
Offers items in Marketplace?	Yes	Yes but requires an offline Marketplace syndication tool
Support for **Active Directory Authentication Library (ADAL)**?	Yes	Yes
Support for tools such as Azure CLI, Visual Studio, and PowerShell?	Yes	Yes
Create service principals via the Azure portal?	Yes	No

Capability or scenario	Azure Active Directory (AAD)	Active Directory Federation Services (ADFS)
Create service principals with certificates?	Yes	Yes
Create service principals with secrets or keys?	Yes	Yes
Applications can use the Graph service?	Yes	No
Applications can use an identity provider for sign-in?	Yes	Yes but requires applications to federate with an on-premises ADFS instance
Managed system identities?	No	No

As discussed earlier in this chapter, the Open ID Connect protocol is used as part of the authorization flow and also the resource owner flow. We utilize the JSON Web Token for use of the Azure portal and use Azure tools to manage both Azure and Azure Stack Hub. Here, we make use of the **Active Directory Authentication Library (ADAL)**.

Let's now dive straight into the directory-based authentication topologies that are supported from within Azure Stack Hub.

Overviewing directory-based authentication in depth

Microsoft Azure Stack Hub supports multiple Active Directory topologies for identity and authentication depending on the identity provider chosen. There are also differences depending on whether this Azure Stack Hub is a single-tenant or multi-tenancy deployment. By default, when Azure Stack Hub is deployed with AAD as the identity provider, then it is configured as a single-tenant topology. Let's start by discussing that topology in more detail.

Understanding the AAD single-tenant topology

This is the default topology that is chosen when you deploy Azure Stack Hub with AAD selected as the identity provider. A single-tenant topology is useful when all users are part of the same tenant. It is the typical topology used by an organization that hosts an Azure Stack Hub instance. This is shown in the following figure:

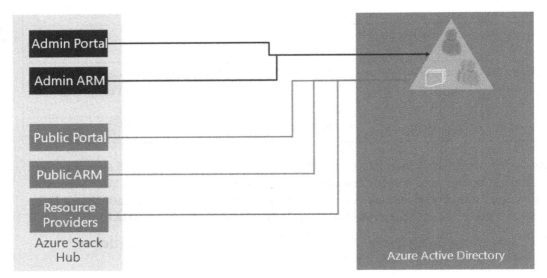

Figure 4.2 – A working of the AAD single-tenant topology

This topology means that Azure Stack Hub registers all applications and services to the same AAD tenant. Additionally, it means that Azure Stack Hub only authenticates users and applications from that tenant directory including tokens. Identities for administrators, cloud operators, and users are all in the same tenant directory. To allow a user from another tenant directory to access this Azure Stack Hub instance, they must be invited as a guest user to the tenant directory.

Understanding the AAD multi-tenancy topology

Azure Stack Hub administrators and cloud operators can configure Azure Stack Hub to allow access to applications by tenants from one or more organizations. Users can access applications through the Azure Stack Hub user portal. With this configuration, the administration portal used by the Azure Stack Hub administrators or cloud operators is limited to users from a single-tenant directory. A multi-tenant topology will typically be used by a service provider who wants to allow users from multiple organizations to access the same Azure Stack Hub instance such as a **Cloud Service Provider** (**CSP**). This topology is shown in the following diagram:

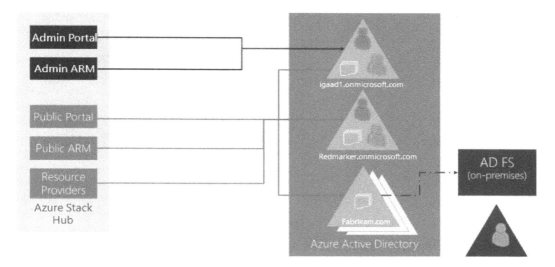

Figure 4.3 – The AAD multi-tenant topology

This topology allows resources to be accessed on a per organization basis. Users within one organization are not able to grant access to resources to users who are outside of their own organization. Identities for Azure Stack Hub administrators or cloud operators can be stored in a different directory tenant from the identities for organizational users. This will provide account isolation at the identity provider level.

Understanding the ADFS topology

The ADFS topology will be used if the Azure Stack Hub instance is deployed in a disconnected scenario. It will also be used if Azure Stack Hub is connected, but only if ADFS was selected as the identity provider. This is shown in the following diagram:

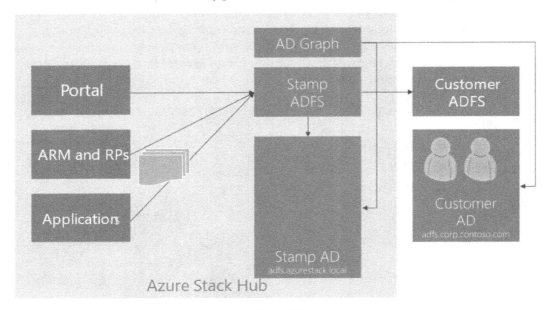

Figure 4.4 – The ADFS topology

The support of this topology within a production environment will require a federation trust to be configured between the built-in Azure Stack Hub ADFS instance and an existing ADFS instance. The built-in ADFS instance within Azure Stack Hub is based on Windows Server 2019. Therefore, the on-premises ADFS and AD instances must be based on Windows 2012 or higher. You can choose to integrate the Graph service from within Azure Stack Hub with your existing AD instance, or, alternatively, you can make use of the OData-based Graph API service, which supports APIs that are consistent with the AAD Graph API. The interactions between the on-premises AD instance and the built-in ADFS are not limited to the OpenID Connect protocol, so they can use any mutually supported protocol. Users are managed and created within the on-premises AD instance, while service principals and the registration of applications are managed from the built-in ADFS instance from Azure Stack Hub. ADFS/AD integration can only happen to a single AD forest that has ADFS deployed and an ADFS trust enabled.

AD Graph is an OData-based web service that supports **create, read, update, delete** (**CRUD**), and query actions against both AD and ADFS instances. It supports a subset of APIs that are required by Azure Stack Hub. It is co-located with ADFS and uses the same contract as AAD Graph. It facilitates the interaction with the Azure Stack Hub fabric **Active Directory Domain Services** (**ADDS**) and customer active directory. It supports the **lightweight directory access protocol** (**LDAP**) protocol for querying the customer's directory. All ADFS operations are performed by PowerShell cmdlets.

Understanding RBAC

RBAC allows an administrator or cloud operator to secure Azure Stack Hub resources with granular permissions. These permissions can be assigned to users, groups, or service principals. Azure Stack Hub comes with some built-in roles, which make it easier to get started and to create custom roles as required. Custom roles that are created in Azure Stack Hub are consistent with Azure, and the people picker experience is integrated into the UI of the portal.

The following is an example of a JSON object that is used for the creation of a custom role:

```
{
    "Name": "Azure Stack Hub registration role",
    "Id": null,
    "IsCustom": true,
    "Description": "Allows access to register Azure Stack Hub",
    "Actions": [
        "Microsoft.Resources/subscriptions/resourceGroups/write",
        "Microsoft.Resources/subscriptions/resourceGroups/read",
        "Microsoft.AzureStack/registrations/*",
        "Microsoft.AzureStack/register/action",
        "Microsoft.Authorization/roleAssignments/read",
        "Microsoft.Authorization/roleAssignments/write",
        "Microsoft.Authorization/roleAssignments/delete",
        "Microsoft.Authorization/permissions/read",
        "Microsoft.Authorization/locks/read",
        "Microsoft.Authorization/locks/write"
    ],
    "NotActions": [
    ],
    "AssignableScopes": [
```

```
        "/subscriptions/<SubscriptionID>"
    ]
}
```

RBAC is used purely for the Azure Stack Hub administration and allows the management of resources in Azure Stack Hub, such as virtual machines, storage, and networks. AAD is not an Azure Stack Hub resource. Roles are composed of actions, scopes, and not actions or excluded operations.

There are two key concepts when it comes to roles. Firstly, there is the role definition, which describes the set of permissions and can be used for multiple assignments. Secondly, there are role assignments. Role assignments associate role definitions with a particular identity (that is, a user or group) at a scope (that is, a resource group). Assignments are always inherited, and subscription assignments will apply to all resources.

There are three built-in roles within Azure Stack Hub. First, there is the **Owner** role, which is allowed to perform all actions within Azure Stack Hub. Second, there is the **Contributor** role, which allows all actions to be performed within Azure Stack Hub with the exception of writing or deleting role assignments. Third, there is the **Reader** role, which only allows read actions within Azure Stack Hub. For those of you that have used SharePoint in the past, these roles might already be familiar to you.

The goal of RBAC is to achieve the principle of least privilege. This is to say that the user has the relevant permission to do their job but no more than that. The best practice is to use either the portal UI or ARM API to create and assign roles. Ensure that the right role is assigned to the relevant identity. Use resource groups to form collections of multiple resources that require the same RBAC permissions:

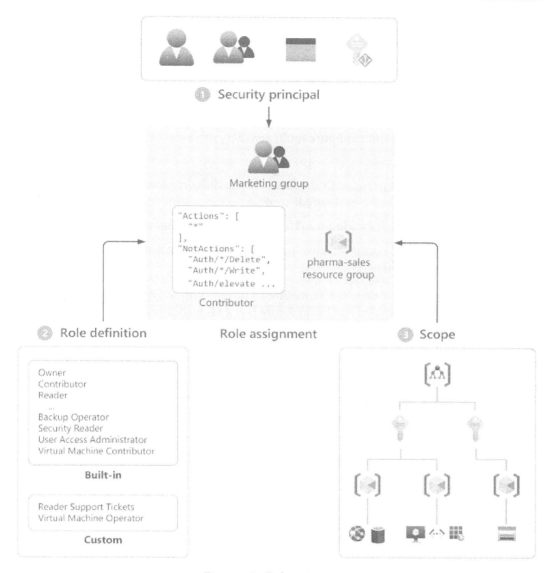

Figure 4.5 – Role assignment

We now understand the different topologies that are supported within Azure Stack Hub and have looked at RBAC in some detail. These are key concepts that are likely to be tested as part of the *AZ-600* exam. Before we move on to the next section, let's take a walk-through to set up Azure Stack Hub to be multi-tenancy.

Configuring multi-tenancy in Azure Stack Hub

It is possible to configure Azure Stack Hub to support users from multiple AAD tenants, thereby allowing them to use services from Azure Stack Hub. It should be noted that after applying some Azure Stack Hub updates, you may need to reconfigure the guest registration. The same might also be required after the installation and update of any new resource providers.

To enable Azure Stack Hub and configure multi-tenancy, there are some prerequisites that must be taken into account:

- The administrative steps for both the Azure Stack Hub directory and the guest directory tenant must be coordinated.
- PowerShell for Azure Stack Hub must be installed and configured.
- Azure Stack Hub tools must be downloaded and an identity module imported into PowerShell.

The following code sample configures Azure Stack Hub to allow a sign-in from a guest directory AD tenant and must be run by an Azure Stack Hub administrator:

```
## Azure Resource Manager Endpoint
$adminARMEndpoint = "https://adminmanagement.local.azurestack.
external"
## Azure Stack Hub directory
$azureStackDirectoryTenant = "northwinds.onmicrosoft.com"
## Guest directory tenant
$guestDirectoryTenantToBeOnboarded = "southwinds.onmicrosoft.
com"
## Name of resource group
$ResourceGroupName = "system.local"
## Region location of resource group
$location = "local"
## Subscription Name
$SubscriptionName = "Default Provider Subscription"
## Configure Azure Resource Manager to accept sign-in from
users in guest domain
Register-AzSGuestDirectoryTenant -AdminResourceManagerEndpoint
$adminARMEndpoint `
  -DirectoryTenantName $azureStackDirectoryTenant `
  -GuestDirectoryTenantName $guestDirectoryTenantToBeOnboarded `
```

```
-Location $location `
-ResourceGroupName $ResourceGroupName `
-SubscriptionName $SubscriptionName
```

Over on the guest AAD, which was southwinds in the preceding example, the administrator must register Azure Stack Hub from northwinds using the same commands:

```
## Azure Resource Manager Endpoint
$tenantARMEndpoint = "https://management.local.azurestack.
external"
## Guest directory
$guestDirectoryTenantName = "southwinds.onmicrosoft.com"
## Register Azure Stack Hub with guest directory
Register-AzSWithMyDirectoryTenant `
  -TenantResourceManagerEndpoint $tenantARMEndpoint `
  -DirectoryTenantName $guestDirectoryTenantName `
  -Verbose
```

Users from the southwinds.onmicrosoft.com domain will now be able to log in to the Azure Stack Hub portal in the northwinds domain via the following URL: https://portal.local.azurestack.external/southwinds.onmicrosoft.com.

For the final section of this chapter, we will now move on to look at the AAD and ADFS and the difference they make to our Azure Stack Hub instance.

Configuring ADFS and Graph integration

By selecting ADFS as the identity provider, identities from an existing Active Directory forest are able to authenticate with resources within Azure Stack Hub. The existing Active Directory forest will need a deployment of ADFS instances to enable the creation of an ADFS federation trust.

Authentication is only one part of identity. To be able to manage RBAC in Azure Stack Hub, the Graph component must also be configured. The Graph component is used to look up the user account in the existing Active Directory forest when access to the resource is delegated. This is done using the LDAP protocol:

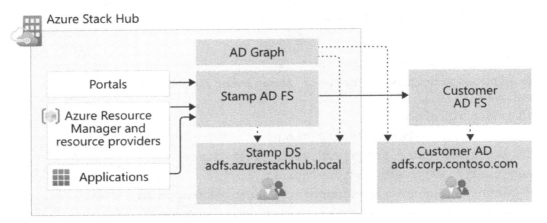

Figure 4.6 – The ADFS Graph topology

The existing ADFS is the account **Security Token Service (STS)**, which sends the claims to Azure Stack Hub ADFS (that is, the resource STS). Automation in Azure Stack Hub creates the claims provider trust with the metadata endpoint for the existing ADFS.

A relying party trust must be configured in the existing ADFS. This must be configured by the operator as it is not performed by the automation. The pattern for the Azure Stack Hub VIP endpoint for ADFS is https://adfs.<region>.<externalFQDN>/.

The relying party trust configuration also means that you have to configure the claim transformation rules that are supplied by Microsoft.

A service account must be provided that has read permission in the existing Active Directory for the Graph configuration. This account is necessary as input for the automation to enable RBAC.

A new owner is also configured for the default provider subscription, which has full access to all resources when signed into the Azure Stack Hub administrator portal.

Graph integration only supports the use of a single Active Directory forest. If multiple forests exist, then only the forest included in the configuration will be used to authenticate users and groups.

The closest Active Directory site to your Azure Stack Hub should be configured. This means you can avoid having the Azure Stack Hub Graph service resolve queries using a Global Catalog server from a remote location.

To trigger automation and configure Graph integration, the following commands should be run from a PowerShell window as an administrator. Then, use the CloudAdmin credential to authenticate using the privileged endpoint:

```
$creds = Get-Credential
$pep = New-PSSession -ComputerName <IP Address>
-ConfigurationName PrivilegedEndpint -Credential $creds
$i = @(
        [pscustomobject]@{
                CustomADGlobalCatalog="fabrikam.com"
                CustomADAdminCredential= get-credential
                SkipRootDomainValidation = $false
                ValidateParameters = $true
        })

  Invoke-Command -Session $pep -ScriptBlock {Register-
DirectoryService -customCatalog $using:i}
```

Following on from the automation of Graph, we need to trigger the automation to configure the claims provider trust in Azure Stack Hub. This can be accomplished with the following commands in the same administrator PowerShell:

```
$creds = Get-Credential
Enter-PSSession -ComputerName <IP Address of ERCS>
-ConfigurationName PrivilegedEndpoint -Credential $creds
Register-CustomAdfs -CustomAdfsName Contoso
-CustomADFSFederationMetadataEndpointUri https://
win-SQOOJN70SGL.contoso.com/federationmetadata/2007-06/
federationmetadata.xml
Set-ServiceAdminOwner -ServiceAdminOwnerUpn administrator@
contoso.com
```

When you rotate the certificate on the existing ADFS, you must set up the ADFS integration again.

Microsoft provides the script that configures the relying party trust, which includes the claim transformation rules. This can be downloaded from the following URL: `https://github.com/Azure/AzureStack-Tools/tree/vnext/` `DatacenterIntegration/Identity`.

This integration is only required if you choose to use ADFS as the identity provider, and this leads us nicely into the next section on deciding between AAD and ADFS.

Deciding between AAD and ADFS

In the previous section, we touched on the different topologies offered by AAD and ADFS, which we are now going to build on in this section. To ensure that the right decision is made when selecting the identity provider, it is important to understand the differences between the two options. It is imperative that you understand the limitations of choosing ADFS as the identity provider whether in a connected or disconnected scenario.

The connected Azure Stack Hub deployment

When you choose a connected deployment of Azure Stack Hub, as stated earlier in this chapter, you can then select either AAD or ADFS as the identity provider. The choice of identity provider has no bearing on tenant virtual machines and the identity provider they can use. The tenant virtual machines can use an identity provider depending on how they are going to be configured. This means that the tenant virtual machines can still choose AAD, Windows Server Active Directory domain-joined, workgroup, and more. This is completely unrelated to the choice of identity provider for Azure Stack Hub.

To use AAD as the identity provider, we need two AAD accounts. The two accounts are as follows:

- The global admin account
- The billing account

These two accounts can either be the same account or different accounts. Many organizations will tend to use the same account to keep things simple. The recommendation and best practice, however, is to utilize separate accounts.

The global admin account is an AAD account, which is used to create applications and service principals for Azure Stack Hub infrastructure services in AAD. This account is required to have directory admin permissions to the directory that the Azure Stack Hub system is going to be deployed into. This account will become the cloud operator global admin for the AAD user and is responsible for the following tasks:

- Provision and delegate applications and service principals for all Azure Stack Hub services that need to interact with AAD or the Graph API.

- It acts as the service administrator account. This is the account that is the owner of the default provider subscription. This account can be used to log into the Azure Stack Hub administrator portal. It can be used to create plans and offers, set quotas, and perform other administrative functions within Azure Stack Hub.

The global administrator account is not required once the deployment is complete. It is not needed to run Azure Stack Hub post-deployment. This account can be disabled or secured according to Microsoft's best practices.

The billing account is used to establish the billing relationship between the Azure Stack Hub solution and the Azure commerce backend. This is the account that is billed for any Azure Stack Hub fees. This is the same account that is used for offering items in Marketplace and other hybrid scenarios.

The disconnected Azure Stack Hub deployment

A disconnected deployment of Azure Stack Hub must use an ADFS identity provider, but it does not restrict you from later connecting the Azure Stack Hub instance to Azure for hybrid tenant virtual machines. Typically, the disconnected deployment is chosen for environments that have security or other restrictions, which means the environment is not connected to the internet. It can also be selected when the organization does not wish to share data, including usage data with Azure. The other typical usage of a disconnected Azure Stack Hub deployment is when it is being used as a private cloud deployed on the corporate network with no hybrid requirements.

Adding an Azure Stack Hub user account in AAD

To be able to test offers and plans or create resources then a user account will need to be created in the AAD tenant. This user account can either be created using the Azure portal or by using PowerShell.

To create the user in the Azure portal, using the left-hand navigation bar, select **Azure Active Directory**, then **Users**, and click on the **New User** button, as shown in the following screenshot:

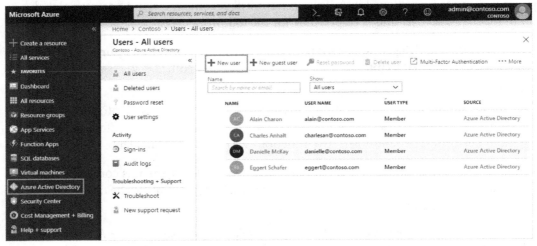

Figure 4.7 – Adding a user via the Azure portal

This will allow you to populate the user page with the following information:

- **Name**: This is the full name of the user including the first and last name.

- **User Name**: This is the username of the new user normally in the format of `name@<domainname>.onmicrosoft.com`.

- **Profile**: This is an optional parameter for adding more information about the user.

- **Directory Role**: The user is the default.

It is important to copy the autogenerated password by checking the show password checkbox and copying the password to the clipboard, as this is needed once the user is created to log into the portal.

Once you click on **Create**, then the new user is created in AAD. It is now possible to log in to the Azure portal with this new account and change the password at `https://portal.local.azurestack.external`.

Adding a user account via PowerShell

If you do not have access to the Azure portal, then it is still possible to add a tenant user account through the use of PowerShell. To be able to use the PowerShell cmdlet to add an **AAD** user, the first step is to download the Microsoft Online Services Sign-In Assistant for IT Professionals RTW from the Microsoft website (`https://www.microsoft.com/en-gb/download/details.aspx?id=28177`).

Once this is installed, then the Microsoft AAD module for PowerShell can be installed, which is used to create the user. The installation of this module can be performed by following these steps:

- Open an elevated Windows PowerShell Command Prompt.

- Run the `Install-Module MSOnline` command in the PowerShell window:

    - If prompted to install the NuGet provider, type Y and press *Enter*.

    - If prompted to install the module from PSGallery, type Y and *Enter*.

Once the module has been added, run the following cmdlets to create the AAD user:

```
$msolcred = get-credential
connect-msolservice -credential $msolcred
$user = new-msoluser -DisplayName "Tenant Admin"
-UserPrincipalName <username>@<yourdomainname> -Password
<password>
        Add-MsolRoleMember -RoleName "Company Administrator"
-RoleMemberType User -RoleMemberObjectId $user.ObjectId
```

As with the previous method of creating an AAD user, it is possible to use the newly created account to log in to the Azure portal at `https://portal.local.azurestack.local`.

Adding new Azure Stack Hub user accounts in ADFS

It is possible to use the Active Directory Users and Computers snap-in to add additional users to an Azure Stack Hub environment, which is using ADFS as the identity provider. The following steps can be used:

1. From a computer connected to the domain and logged in with an account that has access to the Windows Administrative Tools, open a new **Microsoft Management Console** (**MMC**).

2. From the MMC, select **File** and **Add** or **Remove** the snap-in.

3. Select **Active Directory Users** and navigate to **Computers** | `<Directory Domain>` | **Users**.

4. Select **Action** | **New** | **User**.

5. In the **New Object User** tab, provide your user details and click on **Next**.

6. Provide the password and then retype this for confirmation.

7. Select **Next** to complete this value.

8. Select **Finish** to create the user.

The preceding user creation process is exactly the same as a normal domain user, not just an Azure Stack Hub.

Understanding application identity

Applications may need to deploy or configure resources through the ARM, and to do this, they must be represented by their own identity. A user is represented by a security principal called a **user principal**, and an application is represented by a service principal. The service principal is effectively an identity for the application, which allows the delegation of permissions to the application.

For example, there might be a configuration management application that uses ARM to inventory Azure resources. In this scenario, a service principal can be created, which is granted the *Reader* role to that service principal and limits the application to read-only access.

A common use case is Terraform. Terraform is a tool that is used to deploy **Infrastructure as Code (IaC)**. A service principal is used to authenticate Terraform to your Azure Stack Hub to be able to deploy virtual machines, virtual networks, and Network Security Groups. The service principal will need to be assigned a *Contributor*-like role to allow for resource management.

An application must present credentials during authentication. This authentication requires two elements, as follows:

- **Application ID**: This is a **Globally Unique ID (GUID)** that uniquely identifies the application registration in the AD tenant.

- **Secret**: This is either a client secret string or an *X.509* certificate.

It is best practice to run an application under its own identity rather than running it under a user's identity. Some of the reasons for this are listed as follows:

- **Stronger credentials**: An application can make use of *X.509* certificates instead of cleartext username and password.

- **Restrictive permissions**: An application can adhere to the principle of least privilege and only be given the permissions it needs.

- **Permission changes**: An application permission change is less frequent than that of a user permission as users change roles and leave.

An application is registered in the tenant directory, which, in turn, creates the associated service principal object to represent the application identity.

If Azure Stack Hub has been deployed with AAD as the identity provider, then you can create service principals in the same way you do for Azure. That is to say, the registration of your application is done via the Azure portal using the following steps:

1. Sign in to the Azure portal using an Azure account that has access to Azure Stack Hub.

2. Navigate to **Azure Active Directory**, then **App registrations**, and select **New registration**.

3. Provide a name for the application and the appropriate supported account types.

4. Select **Web** as the application type underneath **Redirect URI**.

5. Select **Register**, which will then display the overview page once the application is registered. Make a note of the application ID, as this is required for the application to be included in the code.

6. Select the **Certificates & Secrets** page, and then select new client secret.

7. Specify a description and expiry duration, and then click on **Add**.

8. This displays the value for the client secret, which must then be saved and is used by the application along with the application ID to log in to Azure Stack Hub.

If Azure Stack Hub has been deployed with ADFS as the identity provider, then PowerShell must be used to manage the application identity. This must be performed from an elevated PowerShell window, as we demonstrated in the earlier example of how to create a user.

When creating a certificate credential for use with an application, there are certain requirements that must be met. We have detailed them here:

- The certificate must be issued from a Public Certificate Authority or an Internal Certificate Authority. For a test or development environment, it is possible to use a "self-signed" certificate.

- The cryptographic provider must be set as a Microsoft legacy **Cryptographic Service Provider** (**CSP**) key provider.

- It must be in **Personal Exchange Format** (**PFX**) format owing to the fact that both the public and private keys are needed.

- The Azure Stack Hub infrastructure requires network access to the certificate authority's **Certificate Revocation List** (**CRL**) location as published within the certificate.

Once the certificate has been created and installed, it is possible to use a PowerShell script to register the application and create the associated service principal. Then, you can also use this service principal to sign into Azure.

The PowerShell **Privileged Endpoint** (**PEP**) is a remote PowerShell console that provides just enough capabilities to help you perform a required task, and it comes preconfigured as part of Azure Stack Hub. This endpoint only exposes a restricted set of cmdlets through the use of PowerShell **Just Enough Administration** (**JEA**).

Open an elevated PowerShell session, and run the following commands to register the application and create the associated service principal:

```
$Creds = Get-Credential
$Session = New-PSSession -ComputerName "<PepVm>"
-ConfigurationName PrivilegedEndpoint -Credential $Creds
$Cert = Get-Item "<YourCertificateLocation>"
$SpObject = Invoke-Command -Session $Session -ScriptBlock
{New-GraphApplication -Name "<YourAppName>" -ClientCertificates
$using:cert}
$AzureStackInfo = Invoke-Command -Session $Session -ScriptBlock
{Get-AzureStackStampInformation}
$Session | Remove-PSSession
$ArmEndpoint = $AzureStackInfo.TenantExternalEndpoints.
TenantResourceManager
$GraphAudience = "https://graph." + $AzureStackInfo.
ExternalDomainFQDN + "/"
$TenantID = $AzureStackInfo.AADTenantID
```

```
Add-AzEnvironment -Name "AzureStackUser" -ArmEndpoint
$ArmEndpoint
$SpSignin = Connect-AzAccount -Environment "AzureStackUser" `
-ServicePrincipal `
-CertificateThumbprint $SpObject.Thumbprint `
-ApplicationId $SpObject.ClientId `
-TenantId $TenantID
$SpObject
```

Once these commands have been completed, it will show the registration information for the application including the service principals' credentials. The client ID and ThumbPrint are authenticated and authorized for access to resources managed by ARM.

A sample of the output of the command is shown here:

```
ApplicationIdentifier : S-1-5-21-1512385356-3796245103-
1243299919-1356
ClientId              : 3c87e710-9f91-420b-b009-31fa9e430145
Thumbprint            :
30202C11BE6864437B64CE36C8D988442082A0F1
ApplicationName       : Azurestack-MyApp-c30febe7-1311-4fd8-
9077-3d869db28342
ClientSecret          :
PSComputerName        : azs-ercs01
RunspaceId            : a78c76bb-8cae-4db4-a45a-c1420613e01b
```

Once the service principal has been created for testing, you can create a self-signed certificate and then update the ThumbPrint value of the service principals' credentials to match the certificate just created.

The next section of PowerShell code performs these actions:

```
$Session = New-PSSession -ComputerName "<PepVM>"
-ConfigurationName PrivilegedEndpoint -Credential $Creds
$Cert = Get-Item "<YourCertificateLocation>"

$SpObject = Invoke-Command -Session $Session -ScriptBlock
{Set-GraphApplication -ApplicationIdentifier "<AppIdentifier>"
-ClientCertificates $using:NewCert}
$Session | Remove-PSSession
$SpObject
```

The output, as with the previous set of commands, shows the application registration information. And, as you can see, the `Thumbprint` value has now been updated to reflect the new certificate:

```
ApplicationIdentifier : S-1-5-21-1512385356-3796245103-
1243299919-1356
ClientId              :
Thumbprint            :
AF22EE716909041055A01FE6C6F5C5CDE78948E9
ApplicationName       : Azurestack-MyApp-c30febe7-1311-4fd8-
9077-3d869db28342
ClientSecret          :
PSComputerName        : azs-ercs01
RunspaceId            : a580f894-8f9b-40ee-aa10-77d4d142b4e5
```

When registering an application into a production environment, a self-signed certificate should not be used but rather a public certificate from a known certificate authority.

An alternative to using an *X.509* certificate, as shown earlier, is to make use of the client secret credentials instead. It should be noted, however, that this is inherently less secure than using an *X.509* certificate and is not recommended for use in a production environment. It also requires embedding the secret within the client application source code. So although it might be suitable for development environments, it would then need to be updated to an *X.509* certificate when testing before publishing to production.

I have shown the PowerShell commands to create the service principal with a client secret credential here for completeness:

```
$Creds = Get-Credential
$Session = New-PSSession -ComputerName "<PepVM>"
-ConfigurationName PrivilegedEndpoint -Credential $Creds
$SpObject = Invoke-Command -Session $Session
-ScriptBlock {New-GraphApplication -Name "<YourAppName>"
-GenerateClientSecret}
$AzureStackInfo = Invoke-Command -Session $Session -ScriptBlock
{Get-AzureStackStampInformation}
$Session | Remove-PSSession
$ArmEndpoint = $AzureStackInfo.TenantExternalEndpoints.
TenantResourceManager
$GraphAudience = "https://graph." + $AzureStackInfo.
ExternalDomainFQDN + "/"
```

```
$TenantID = $AzureStackInfo.AADTenantID
Add-AzEnvironment -Name "AzureStackUser" -ArmEndpoint
$ArmEndpoint
$securePassword = $SpObject.ClientSecret | ConvertTo-
SecureString -AsPlainText -Force
$credential = New-Object -TypeName System.Management.
Automation.PSCredential -ArgumentList $SpObject.ClientId,
$securePassword
$SpSignin = Connect-AzAccount -Environment "AzureStackUser"
-ServicePrincipal -Credential $credential -TenantId $TenantID
$SpObject
```

As with the preceding commands, the last line of this script displays the application registration information, which is shown next:

```
ApplicationIdentifier : S-1-5-21-1634563105-1224503876-
2692824315-2623
ClientId              : 8e0ffd12-26c8-4178-a74b-f26bd28db601
Thumbprint            :
ApplicationName       : Azurestack-YourApp-6967581b-497e-4f5a-
87b5-0c8d01a9f146
ClientSecret          : 6RUWLRoBw3EebBLgaWGiowCkoko5_j_
ujIPjA8dS
PSComputerName        : azs-ercs01
RunspaceId            : 286daaa1-c9a6-4176-a1a8-03f543f90998
```

As you can see from the information displayed, this does not use an *X.509* certificate, as the Thumbprint value is empty. Additionally, the ClientSecret parameter contains a value, which, in the preceding example, it did not. The ClientSecret parameter is shown only once and, if lost, cannot be retrieved. It would require the secret to be regenerated. The command to regenerate the secret is similar to the preceding command with a couple of slight changes, as I have shown here:

```
# Create a PSSession to the PrivilegedEndpoint VM
$Session = New-PSSession -ComputerName "<PepVM>"
-ConfigurationName PrivilegedEndpoint -Credential $Creds


# Use the privileged endpoint to update the client secret, used
by the service principal associated with <AppIdentifier>
$SpObject = Invoke-Command -Session $Session -ScriptBlock
{Set-GraphApplication -ApplicationIdentifier "<AppIdentifier>"
```

```
-ResetClientSecret}
$Session | Remove-PSSession


# Output the updated service principal details
$SpObject
```

Before we move on to look at assigning roles, let's continue with the application registration and quickly run through the removal of an application registration and the associated service principal. The following PowerShell cmdlets can be used to remove an application registration:

```
# Sign in to PowerShell interactively, using credentials
that have access to the VM running the Privileged Endpoint
(typically <domain>\cloudadmin)
$Creds = Get-Credential


# Create a PSSession to the PrivilegedEndpoint VM
$Session = New-PSSession -ComputerName "<PepVM>"
-ConfigurationName PrivilegedEndpoint -Credential $Creds


# OPTIONAL: Use the privileged endpoint to get a list of
applications registered in AD FS
$AppList = Invoke-Command -Session $Session -ScriptBlock
{Get-GraphApplication}


# Use the privileged endpoint to remove the application and
associated service principal object for <AppIdentifier>
Invoke-Command -Session $Session -ScriptBlock {Remove-
GraphApplication -ApplicationIdentifier "<AppIdentifier>"}
```

This cmdlet returns no output, but you will see verbatim confirmation output on the console while the cmdlet is running.

Assigning a role

Earlier in this chapter, we looked at RBAC, and as with users, access to resources for applications is controlled by RBAC, so the application service principal must have a role assigned to be able to use resources from Azure Stack Hub.

The type of resource chosen establishes the access scope for the application. This scope can be set at the subscription, resource group, or resource level. By default, permissions are inherited at the lower levels of scope. In an ADFS deployment, if you wish to add a group as a principal in an RBAC role, then the group must be of the Security Universal group type in AD and not a Global or Local group.

To be able to add a role assignment to a given resource for a specific service principal, the user account being used must have the `Microsoft.Authorisation/roleAssignment/write` permission either directly or as part of their role assignment. The **Owner** and **User Access Administrator** built-in roles have this permission.

To assign an application service principal to a role at the subscription scope, inside the Azure Stack Hub user portal, select **Subscriptions** and the relevant subscription:

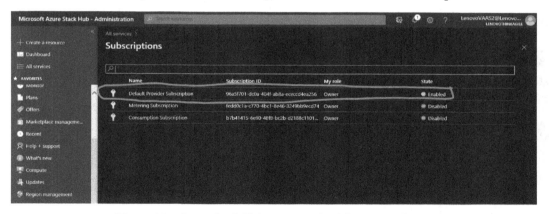

Figure 4.8 – Azure Stack Hub user portal – Subscription Scope

Select the **Access control (IAM)** page and then select **+Add**. Pick the role required for the application in **Role**, and then search for the application in **Select**. Once this has been found, select the relevant application and clock on **Save** to finish the role assignment. This process is shown in the following screenshot:

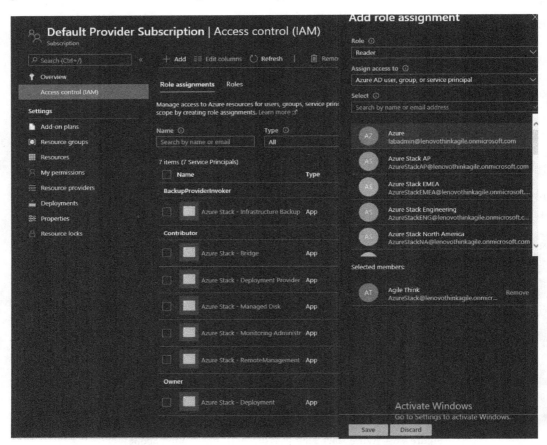

Figure 4.9 – The Azure Stack Hub portal – assigning a role

This has now given the application an identity and has authorized it to access Azure Stack Hub resources. This application can now be enabled to sign in and securely access Azure Stack Hub resources. This completes our coverage of the identity model in this chapter. However, before we move on, let's review a quick recap on what we have learned.

Summary

This chapter has given us a great insight into the Azure Stack Hub identity model, which is key learning for the *AZ-600* exam. We started the chapter with some background and common terminology that we need to understand. From there, we dived into understanding the Azure identity model. This introduced us to the different layers that identity is used for and the various authentication processes we need to be aware of. We were introduced to different token types, such as the JSON Web Token and the access token. We walked through the authentication flow for the Azure Stack Hub portal. We then looked through the different identity providers supported by Azure Stack Hub, AAD, and ADFS. We covered the limitations of choosing ADFS as an identity provider.

Then, we moved on to the AD topologies supported by identity in Azure Stack Hub. We covered the implementation of RBAC and how permissions could be assigned to users in the different AD topologies to access resources in Azure Stack Hub. We are now able to describe the three built-in roles within Azure Stack Hub: Owner, Contributor, and Reader. This understanding is key to some of the concepts of the *AZ-600 exam*, including multi-tenancy.

We finally walked through AAD and ADFS. We touched on the differences between connected and disconnected deployment scenarios for Azure Stack Hub, building on what we have learned in previous chapters. We discussed the different techniques to add users to the Azure Stack Hub identity provider. We then went into detail about application identity. We walked through the registration of an application and the creation of the associated service principal. We then finished the chapter by returning to RBAC and looking at the assignment of a role to an application, which allows it to access resources within Azure Stack Hub.

From identity, we are now going to move on to security, in the next chapter; so, we will be moving from authentication to authorization.

5
Securing Your Azure Stack Hub Instance

Following on from the previous chapter where we covered identity, the next logical step is to move on and delve into security. In this chapter, we are going to focus on the infrastructure security available within **Microsoft Azure Stack Hub**. We will also cover how Microsoft Azure Stack Hub compliance is managed and the tools that can be used to ensure that security is maintained within Microsoft Azure Stack Hub. We will cover security best practices within Microsoft Azure Stack Hub to ensure that we limit the attack surface available to hackers. We will also cover tools for investigating breaches. This chapter will help you understand and describe the approach that Microsoft takes toward security within Azure Stack Hub.

The following sections will provide coverage of these topics:

- Overview of infrastructure security
- Understanding the benefits of a locked system
- Reviewing Azure Stack Hub integrated compliance
- Approaching security for Azure Stack Hub

So, let's begin with an overview of infrastructure security.

Technical requirements

You can view this chapter's code in action here: `https://bit.ly/3DhSEzV`

Overview of infrastructure security

The traditional approach to cybersecurity used to be *build a bigger wall and a bigger moat*, but this did not always keep people out. Unfortunately, in this day and age, if someone really wants to get into a system, they are likely to get in. When working with Microsoft, there are two fundamental things they will tell you:

- You are in the fight, whether you thought you were or not.

- You have almost certainly already been penetrated.

This is not meant to be a scare tactic from Microsoft. but more a view on the reality of systems today. That having been said, Microsoft does have a clear vision on security for Azure Stack Hub. This is defined by Microsoft with the following statement:

> *Provide our customers with a similar level of security and protection to Azure, while running Azure Stack Hub in hostile environments.*

So how do they achieve this?

Microsoft has two fundamental principles when it comes to securing Azure Stack Hub. Firstly, Azure Stack Hub is hardened by default and, when running Azure Stack Hub, you should always assume breach. We will look at these in detail as we work through this chapter. Azure Stack Hub effectively has two security posture levels that coexist. The first layer is the Azure Stack Hub infrastructure layer, which includes the hardware components up to and including **Azure Resource Manager** (**ARM**). The second layer is for the workloads created, deployed, and managed by the tenants. The security of the tenant workloads is the responsibility of users, which is consistent with Azure. Both of these are shown in the following diagram:

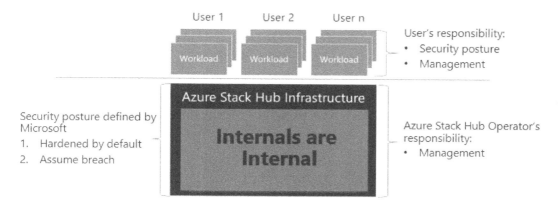

Figure 5.1 – Azure Stack Hub security responsibilities

As shown in the diagram and mentioned earlier, the security posture of Microsoft is hardened by default and assume breach. Let's take a look at the first of these – hardened by default.

Hardened by default

The concept of hardened by default is to ensure that Azure Stack Hub is production-ready, and that security is not an afterthought that needs to be configured step by step.

This concept starts from the hardware that is certified by Microsoft and supplied by the **original equipment manufacturer** (**OEM**) vendors, including **Lenovo**, **Dell**, **HPE**, and **Cisco**. We effectively create a hardware root of trust and all OEM vendors must ensure that any solution they offer for Azure Stack Hub includes the following components:

- Secure boot

- **Unified Extensible Firmware Interface** (**UEFI**)

- TPM 2.0

Azure Stack Hub dictates secure boot on all Hyper-V hosts and infrastructure virtual machines. As Azure Stack Hub is built on an integrated system, the security posture is defined by Microsoft and not by the OEM vendor.

The idea behind the hardened by default approach is to try and limit the software attack surface. To achieve this, Microsoft has built Azure Stack Hub on a hardened security OS baseline, which is based on DISA STIG. **DISA** is the **Defense Information Systems Agency** and they provide guidance and configuration standards or **Security Technical Implementation Guides (STIGs)** for the technical lockdown of information systems.

Azure Stack Hub also makes use of Windows Server 2019 security features, including the following:

- Code Integrity/Windows Defender Application Control (Device Guard)
- Antimalware (Windows Defender)

Windows Defender Application Control (formerly Code Integrity in Windows 2016) allows the filtering of executables and ensures that only authorized code is allowed to run on Azure Stack Hub. Authorized code must be signed either by Microsoft or the OEM vendor who supplied the Azure Stack Hub integrated system. Microsoft defined a policy that includes a list of authorized software, including software from all of the OEM vendors. Only software that is listed in this policy is allowed to run on Azure Stack Hub. Any attempt to run or execute software that is not authorized will be blocked and an alert raised.

Each component of Azure Stack Hub, including Hyper-V hosts and virtual machines, are protected with Windows Defender antivirus.

When Azure Stack Hub is deployed in a connected state, antivirus definition and engine updates are applied automatically multiple times a day. When deployed, disconnected antimalware updates are applied as part of the monthly Azure Stack Hub updates. It is possible to import Windows Defender updates to be able to update disconnected Azure Stack Hub more frequently.

Alongside these security features, the attack surface is further reduced by removing unused components and disabling legacy protocols such as **Secure Sockets Layer (SSL)** and **server message block (SMBv1)**.

Azure Stack Hub infrastructure includes multiple layers of network **access control lists (ACLs)**, which help to prevent unauthorized access to the infrastructure components and also limit the communication of these components to just the paths that are required to function correctly. These network ACLs are enforced in the following layers of Azure Stack Hub:

- Top-of-rack switches (Layer 1)
- Software-defined networking (Layer 2)
- Guest firewalls (Layer 3)

Azure Stack Hub also uses security controls around secrets and sensitive data as part of this hardened by default approach.

Data at rest is protected by hardware-rooted data at rest encryption using BitLocker and **Trusted Platform Module** (**TPM**) 2.0. This encryption is used to protect against physical loss or theft of Azure Stack Hub storage components.

Azure Stack Hub infrastructure components also protect data in transit by only communicating using channels encrypted with TLS 1.2. The certificates used for this encryption are self-managed by the infrastructure and never seen by administrators. In addition to these internal communication channels, the external infrastructure endpoints, such as REST endpoints or the Azure Stack Hub portal, can also be protected using TLS 1.2 for secure communication.

There is also a strong authentication model at play between the infrastructure components through the use of Kerberos, which relies on an *invisible* internal **active directory** (**AD**).

Finally, there is an automated rotation of secrets and certificates that rotates every 24 hours. Azure Stack Hub utilizes multiple secrets, such as passwords and certificates, to function. All internal service accounts have their associated passwords rotated as part of this process as they are associated with a special type of domain account that is managed directly by the internal domain controller. All internal certificates within Azure Stack Hub use 4,096-bit RSA keys.

This covers the features for hardened by default, so let's move on to assume breach.

Assume breach

Assume breach means devoting resources to detecting and limiting the impact of breaches of assets rather than only trying to prevent attacks.

Part of this restriction is due to the constrained operator experience. Azure Stack Hub utilizes operators rather than administrators and there are no domain administrator credentials required to administer Azure Stack Hub. The administrator role is the most targeted in any organization and if this is breached, then attackers have the keys to the kingdom.

As part of this constrained operator experience, there are limited interaction points exposed and each of these has a distinct purpose:

- **Administrator portal**: Provides the point-and-click user experience for daily operations management
- **ARM**: Provides a REST API for all management operations and is used by PowerShell and the Azure **command-line interface (CLI)**
- **Privileged endpoint (PEP)**: A PowerShell endpoint that exposes a limited set of PowerShell cmdlets that are heavily audited

Azure Stack Hub adheres to some security best practices during development as it is covered by the Microsoft Secure Development Lifecycle and is regularly penetration tested by third-party organizations. In addition to this, the threat model of each component is considered during development. Patching and updating of the entire infrastructure in Azure Stack Hub is automated, including security patches. Finally, as mentioned previously, the secrets and passwords used internally in Azure Stack Hub are rotated regularly.

Now that we have an overview of the infrastructure security within Azure Stack Hub and how it controls the attack space, let's turn our attention to the benefits of a closed system, which is our next topic.

Understanding the benefits of a locked system

The benefits of a locked system start from the hardware used to deploy Azure Stack Hub. As we have learned in previous chapters, the Azure Stack Hub solution provided by OEM vendors such as Lenovo, Dell, HPE, and Cisco is an integrated system. This means that the hardware and software is known at deployment time. This gives us a few security advantages, which are detailed as follows:

- **List of software components**: Applications are whitelisted, and Device Guard ensures that only Microsoft-signed software is deployed.
- **OS dependencies**: Azure Stack Hub includes a customized OS configuration with unnecessary legacy applications removed.
- **Known hardware characteristics**: All OEM vendors have data at rest enabled by default.

These properties are the same regardless of which OEM vendor is chosen by the organization and all integrated systems are certified by Microsoft.

The next property that affords a security benefit in this locked system ecosystem is the known **network communications**. Again, all OEM vendors must conform to Microsoft guidance for hosts, VMs, IPs, ports, and protocols. This allows for network whitelisting through the use of ACLs in multiple layers, as we described in the previous section. BitLocker is also enabled by default across the fabric.

As we mentioned in the *Assume breach* section earlier in this chapter, the attack surface is reduced by the constrained administrator experience. This means that there is no unconstrained administrator access and each administrator role, such as the storage admin, adheres to the **just enough administration** (JEA) practice. As with other roles, the method is least privilege, and this includes for administration roles. Where there is a need for elevated administration access, then **just in time** (JIT) administration is provided by the PEP, which is generally used for support scenarios.

The final property that improves the security of Azure Stack Hub is the fact that there is known behavior expected of the platform. That is to say that there are known correct behaviors, potentially known suspect behaviors, and who accesses what is fully audited. This means that all events are aggregated and audited, allowing an operator to see clearly precisely who is using the system and what they are doing. This allows the operator to identify quickly any behaviors out of the ordinary that may potentially be a breach.

Having walked through the benefits of a locked system and reviewed the security advantages this equates to, it is time to move on and review security compliance within the Azure Stack Hub integrated system.

Reviewing Azure Stack Hub integrated compliance

Azure Stack Hub is pre-validated for compliance standards by a formal assessment of the Azure Stack Hub infrastructure, which was conducted by 3PAO.

Azure Stack Hub comes with pre-compiled documentation that applies to controls in the following areas.

- Technology
- Infrastructure

It does not cover tenant applications, people, or processes. The assessments will support getting the required certification for the following compliance regulations:

- **Payment Card Industry Data Security Standard (PCI-DSS)**: Covers the payment card industry
- **FedRAMP**: Covers government customers
- **Cloud Control Matrix (CCM)**: Deals with comprehensive mapping across multiple standards, including the following:

  - ISO 27001

  - **Health Insurance Portability and Accountability Act (HIPAA)**

  - HITRUST

  - ITAR

  - NIST SP800-53

It should be noted that Azure Stack Hub is assessed and not certified for compliance, and further information is available from `https://aka.ms/AzureStackCompliance`, which can be accessed using the Azure Stack Hub portal login credentials.

The FedRamp assessment was conducted by Coalfire between April 2018 and June 2018. The findings in relation to this assessment were that Azure Stack Hub could be effectively used within a FedRamp high-impact-level authorization boundary.

This covers security compliance of the Azure Stack Hub infrastructure, so we will now turn our attention to the approach for configuring security within Azure Stack Hub in the next section.

Approaching security for Azure Stack Hub

As we have previously stated in this chapter, Azure Stack Hub comes hardened by default and production-ready, but there are still some changes that can be made in Azure Stack Hub regarding security. This section will run through some of this security configuration.

When approaching security for Azure Stack Hub, there are several components that should be focused on. These include areas concerning workload protection and associated controls. We should be focusing on creating custom roles for use in the **role-based access control (RBAC)**, as discussed in *Chapter 4, Exploring Azure Stack Hub Identity*.

We should also be focusing on our virtual networking security and ensuring that we review the security groups and egress security, especially when it comes to software-defined networking. To cover networking in a little more detail, let's take a look at a particular component – **Transport Layer Security (TLS)**.

Working with TLS policy

TLS is a widely adopted cryptographic protocol that is used to establish encrypted communication over the network. There have been multiple versions of TLS over time, and Azure Stack Hub infrastructure makes use of TLS 1.2 for all of its communication. The external interfaces exposed by Azure Stack Hub also default to require TLS 1.2. For backward compatibility, however, it is able to negotiate down to TLS 1.1 or even TLS 1.0. This downward negotiation is controlled by the client. Azure Stack Hub will honor the request from the client to negotiate to TLS 1.1 or 1.0 when requested by the client. If no downward negotiation is sent from the client, then Azure Stack Hub will default back to TLS 1.2.

With later releases of Azure Stack Hub, which have introduced the ability to configure the TLS policy, this means that as an Azure Stack Hub operator, you can enforce all communications with Azure Stack Hub to be conducted over TLS 1.2 and remove the backward compatibility for requests from clients running TLS 1.1 or TLS 1.0. This is the recommended approach from Microsoft for production environments running Azure Stack Hub.

The TLS policy can be controlled through PowerShell cmdlets in the PEP, and I have included examples here for reference.

To get the current TLS policy, execute the following command:

```
Get-TLSPolicy
```

This will return the current TLS policy in a format such as TLS_1.2.

To set the TLS policy, execute the following command:

```
Set-TLSPolicy -Version <String>
```

The string for the version is either TLS_1.2, which forces all connections to be TLS version 1.2, or TLS_All, which allows backward compatibility down to TLS 1.1 or TLS 1.0. The update to the TLS policy can take several minutes to apply depending on the environment.

As an example, the following cmdlet will enforce the TLS 1.2 version policy and the corresponding output upon successful completion:

```
Set-TLSPolicy -Version TLS_1.2
VERBOSE: Successfully setting enforce TLS 1.2 to True
VERBOSE: Invoking action plan to update GPOs
VERBOSE: Create Client for execution of action plan
VERBOSE: Start action plan
VERBOSE: Verifying TLS Policy
VERBOSE: Get GPO TLS protocols registry 'enabled' values
VERBOSE: GPO TLS applied with the following preferences:
VERBOSE:    TLS protocol SSL 2.0 enabled value: 0
VERBOSE:    TLS protocol SSL 3.0 enabled value: 0
VERBOSE:    TLS protocol TLS 1.0 enabled value: 0
VERBOSE:    TLS protocol TLS 1.1 enabled value: 0
VERBOSE:    TLS protocol TLS 1.2 enabled value: 1
VERBOSE: TLS 1.2 is enforced
```

Compare this to the TLS_All switch:

```
Set-TLSPolicy -Version TLS_All
VERBOSE: Successfully setting enforce TLS 1.2 to False
VERBOSE: Invoking action plan to update GPOs
VERBOSE: Create Client for execution of action plan
VERBOSE: Start action plan
VERBOSE: Verifying TLS Policy
VERBOSE: Get GPO TLS protocols registry 'enabled' values
VERBOSE: GPO TLS applied with the following preferences:
VERBOSE:    TLS protocol SSL 2.0 enabled value: 0
VERBOSE:    TLS protocol SSL 3.0 enabled value: 0
VERBOSE:    TLS protocol TLS 1.0 enabled value: 1
VERBOSE:    TLS protocol TLS 1.1 enabled value: 1
VERBOSE:    TLS protocol TLS 1.2 enabled value: 1
VERBOSE: TLS 1.2 is not enforced
```

Now that we have learned how to enforce the use of TLS 1.2 for network communication, let's take a look at another component we mentioned earlier in the chapter – data encryption.

Working with data at rest encryption in Azure Stack Hub

We talked about data at rest encryption when looking at the hardened by default security posture earlier on in the chapter. By default, Azure Stack Hub protects users and infrastructure at the storage subsystem layer by using encryption at rest, which is encrypted by BitLocker with 128-bit AES encryption. The BitLocker keys are kept and persisted within an internal secret store. It is possible to change the AES encryption for BitLocker to 256-bit, but this must be done at the time of deployment.

Many major compliance standards require data at rest encryption as part of their certification. Examples of these include PCI-DSS, FedRamp, and HIPAA. There are no additional configuration steps or extra work required in Azure Stack Hub to meet these requirements.

Data at rest encryption is all about protecting the data being accessed should the physical hard drives from the system be stolen. The data at rest encryption does not, however, protect the data while it is in transit over the network or while it is being used in memory. Extra security is required to protect this data.

Working with BitLocker recovery keys

BitLocker keys for data at rest in Azure Stack Hub are internally managed. This means that they are not required for normal day-to-day operations of the Azure Stack Hub environment. There are some instances of support that may require the BitLocker keys to bring the Azure Stack Hub system back online after a failure.

BitLocker recovery keys should be retrieved and stored in a secure location outside of Azure Stack Hub, such as in a password vault such as KeePass, HasiCorp Vault, or other third-party vault solutions. If these BitLocker recovery keys are not available in certain support scenarios, then there is a potential for data loss, or the requirement of a system restore from a backup.

The BitLocker recovery keys can be retrieved from the Azure Stack Hub system through the use of PowerShell cmdlets. An example of this cmdlet is shown for reference:

```
Get-AzsRecoveryKeys -raw
```

This cmdlet will return the data mapping between the recovery key, computer name, and password ID(s) for each encrypted volume. BitLocker keys will need to be retrieved following each Azure Stack Hub update, which we will cover in the next chapter.

The next topic to cover in this security section is the rotation of secrets.

Rotating secrets in Azure Stack Hub

To help maintain secure communication with infrastructure resources and services within Azure Stack Hub, secrets are used. To ensure the integrity of the Azure Stack Hub infrastructure, it is important to be able to rotate these secrets frequently, and this is dictated by the Azure Stack Hub operators defined by the organizational requirements.

Administrators are made aware when secrets are expiring by means of alerts in the administrator portal. Being able to rotate secrets will help to resolve the following alerts:

- Pending service account password expiration
- Pending internal certificate expiration
- Pending external certificate expiration

These alerts are raised twice. The first alerts are raised at 90 days before expiration, and are raised as warnings. The second alerts are raised at 30 days and at this point, are raised as critical. Operators and administrators need to ensure that the relevant secrets are rotated to resolve these alerts when they are received, as failing to do so may result in the loss of workloads or even the need for a redeployment of Azure Stack Hub.

There are both internal- and external-facing certificates that must be rotated. For the external-facing certificates, new certificates must be generated as per the process defined in *Chapter 3, Azure Stack Hub Deployment*. It should be noted that the rotation of external-facing certificates does not rotate any additional resource provider certificates as an event hub, and that the rotation of these certificates must follow the guidance for the particular resource provider in question.

Certificates are used to protect external-facing services and are provided by the Azure Stack operator covering the following services:

- Administrator portal
- Public portal
- Administrator ARM
- Global ARM
- Administrator Key Vault
- Key Vault
- Admin extension host
- ACS (including blob, table, and queue storage)
- ADFS
- Graph

It is important to validate that the system is healthy before you attempt to rotate secrets. This can be done using the following PowerShell command:

```
Test-AzureStack -group SecretRotationReadiness
```

The new certificates should be prepared and saved to a fileshare, which can be accessed from the ERCS VMs and is accessible by the CloudAdmin identity.

Download the `CertDirectoryMaker.ps1` file from `https://www.aka.ms/azssecretrotationhelper` and save to the fileshare that was just created.

This script can then be executed in PowerShell, which will then create a folder structure under the fileshare, which is dependent on the identity provider used within Azure Stack Hub. An example of the folder structure created for the **Azure Active Directory** (**AAD**) identity provider is shown here for reference:

```
<ShareName>
    Certificates
        AAD
            ACSBlob
                <certname>.pfx
            ACSQueue
                <certname>.pfx
            ACSTable
                <certname>.pfx
            Admin Extension Host
                <certname>.pfx
            Admin Portal
                <certname>.pfx
            ARM Admin
                <certname>.pfx
            ARM Public
                <certname>.pfx
            KeyVault
                <certname>.pfx
            KeyVaultInternal
                <certname>.pfx
            Public Extension Host
                <certname>.pfx
            Public Portal
                <certname>.pfx
```

The replacement certificates are transferred to the `Certificates\AAD` directory as previously or `Certificates\ADFS` if ADFS is the `Identity` provider.

`<certname>` should be in `cert.<regionname>.<externalFQDN>` format corresponding to your Azure Stack Hub deployment.

The following code is executed to rotate the certificates:

```
# Create a PEP Session
winrm s winrm/config/client '@{TrustedHosts= "<IP_address_of_
ERCS>"}'
$PEPCreds = Get-Credential
$PEPSession = New-PSSession -ComputerName <IP_address_of_
ERCS_Machine> -Credential $PEPCreds -ConfigurationName
"PrivilegedEndpoint"


# Run Secret Rotation
$CertPassword = ConvertTo-SecureString "<Cert_Password>"
-AsPlainText -Force
$CertShareCreds = Get-Credential
$CertSharePath = "<Network_Path_Of_CertShare>"
Invoke-Command -Session $PEPSession -ScriptBlock {
    Start-SecretRotation -PfxFilesPath $using:CertSharePath
-PathAccessCredential $using:CertShareCreds
-CertificatePassword $using:CertPassword
}
Remove-PSSession -Session $PEPSession
```

This PowerShell scripts runs against a **PEP**. This is performed in a remote session on the virtual machine that is hosting the PEP. For an integrated system as we described earlier in the book, there are three such virtual machines running these privileged endpoints. These are Azs-ERCS01, Azs-ERCS02, and Azs-ERCS03, which will all be running on different hosts within the Azure Stack Hub system.

This PowerShell script performs the following actions and should take about an hour to complete the rotation of the certificates:

1. It creates a PowerShell session against the PEP as a CloudAdmin.

2. It runs `Invoke-Command`, passing the session ID for the PowerShell session.

3. It runs `Start-SecretRotation` in the session.

If the rotation is successful, then the PowerShell console should display `ActionPlanInstanceID...CurrentStatus: Completed`.

If rotation fails for any reason, then the error message should inform you why and the necessary actions to take. Make the relevant changes as per the error message and then rerun the PowerShell script with `Start-SecretRotation -ReRun`.

This is the process for rotating the external secrets and certificates, which will need to be run on a regular basis as the certificates expire.

The rotation of internal secrets should not need to be conducted unless a breach is suspected, but for some updates from Microsoft, you will be asked to perform this rotation. The process for rotating internal secrets is almost the same as the external rotation and is shown here for reference:

```
# Create a PEP Session
winrm s winrm/config/client '@{TrustedHosts= "<IP_address_of_
ERCS>"}'
$PEPCreds = Get-Credential
$PEPSession = New-PSSession -ComputerName <IP_address_of_
ERCS_Machine> -Credential $PEPCreds -ConfigurationName
"PrivilegedEndpoint"

# Run Secret Rotation
Invoke-Command -Session $PEPSession -ScriptBlock {
    Start-SecretRotation -Internal
}
Remove-PSSession -Session $PEPSession
```

You will see here that the `Internal` parameter is added to the `Start-SecretRotation` command.

As with the external secrets rotation, if this fails, then make the relevant changes as per the error message and rerun the script with `Start-SecretRotation -Internal -ReRun`.

This completes our coverage of the rotation of secrets within Azure Stack Hub, and we will now move on to the next section, which covers the updating of credentials for the **baseboard management controller (BMC)**.

Updating BMC credentials

The BMC is used to monitor the physical state of the servers, orchestrate updates, and perform the maintenance of nodes. The username and password for the BMC is generally updatable by using management applications from the OEM vendor, although it is not necessarily required to update this information on the physical servers first prior to updating in Azure Stack Hub.

As with the rotation section we have just walked through, the BMC is updated through the use of PowerShell against a PEP.

The following example script will prompt for the credentials for the BMC:

```
# Interactive Version
$PEPIp = "<Privileged Endpoint IP or Name>" # You can also use
the machine name instead of IP here.
$PEPCreds = Get-Credential "<Domain>\CloudAdmin" -Message "PEP
Credentials"
$NewBmcPwd = Read-Host -Prompt "Enter New BMC password"
-AsSecureString
$NewBmcUser = Read-Host -Prompt "Enter New BMC user name"


$PEPSession = New-PSSession -ComputerName $PEPIp -Credential
$PEPCreds -ConfigurationName "PrivilegedEndpoint"


Invoke-Command -Session $PEPSession -ScriptBlock {
    # Parameter BmcPassword is mandatory, while the BmcUser
parameter is optional.
    Set-BmcCredential -BmcPassword $using:NewBmcPwd -BmcUser
$using:NewBmcUser
}
Remove-PSSession -Session $PEPSession
```

This is the recommended approach for updating the BMC credentials with Azure Stack Hub.

Another area we will look at in this chapter on security for Azure Stack Hub is the security relating to Azure Stack Hub logs and how Microsoft deals with customer data.

Handling customer data and Azure Stack Hub logs

Microsoft is considered the data controller purely for data that is shared with them from Azure Stack Hub through the use of diagnostics, telemetry, and billing. The customer is the data controller for all other data as Azure Stack Hub is run on their premises.

All data that is passed to Microsoft is encrypted, and there are data access controls in place when Microsoft employees need to access the data to investigate issues raised by the customer. Employees are only given read-only access to this data and all access to the data is carefully logged and audited.

Data is only kept by Microsoft for a maximum of 90 days once a case has been resolved by Microsoft. A customer can request that any data held by Microsoft be deleted before the 90 days has elapsed. If Microsoft needs to share this data with an OEM vendor in order to resolve an issue, then this is only done at the discretion of the customer.

Microsoft supplies scripts and tools for their Azure Stack Hub customers that can be used to collect and upload diagnostics logs to support in order to assist with their troubleshooting investigations. This data is transferred over an HTTPS protected and encrypted connection to Microsoft. These logs are stored, encrypted, and deleted after 90 days once the investigation is complete.

When Azure Stack Hub is deployed in the connected scenario, then it will automatically send telemetry system data to Microsoft. The Azure Stack Hub operator can control what telemetry data is shared with Microsoft. Sensitive data, such as email addresses, usernames, passwords, and credit card details, are never shared as part of this telemetry data.

The only other area of data that is automatically sent to Microsoft is the usage information, which is used for billing. The configuration of this usage information is again controlled by the Azure Stack operator and can be turned off for those who have chosen the capacity model for their billing rather than the pay-as-you-use billing model.

The final section of this security chapter covers the auditing and intrusion detection capabilities found within Azure Stack Hub.

Understanding auditing and intrusion detection logging

All activity that is undertaken on Azure Stack Hub is logged. This includes operations performed by the cloud operator and access by users. These logs can be collected through integration with a **Security Information and Event Management (SIEM)** tool with the Syslog client of Azure Stack Hub. Syslog for Azure Stack Hub supports the common event format payload, which allows this integration. This integration can be used to connect logs from infrastructure events, as shown in the following diagram:

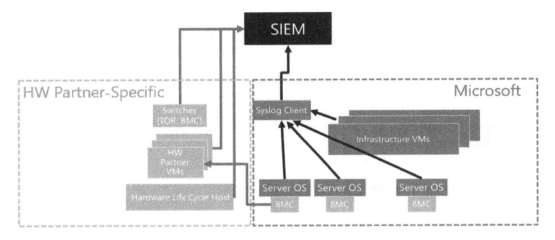

Figure 5.2 – SIEM integration for Azure Stack Hub infrastructure

This diagram shows that information from the servers running Azure Stack Hub and their own internal infrastructure virtual machines can all be connected to Syslog, which in turn connects to the SIEM tool.

Azure Stack Hub has an integrated Syslog client that can be configured to emit messages in the **Common Event Format (CEF)** for their payload.

The Syslog client needs to be configured to forward requests, which can be achieved through the use of PowerShell against the PEP with the following cmdlets. The cmdlets shown as follows will configure Syslog forwarding with TCP, mutual authentication, and TLS 1.2 encryption, which is the recommended approach from Microsoft:

```
# Configure the server
Set-SyslogServer -ServerName <FQDN or ip address of syslog
server> -ServerPort <Port number on which the syslog server is
listening on>
```

```
# Provide certificate to the client to authenticate against the
server
Set-SyslogClient -pfxBinary <Byte[] of pfx file> -CertPassword
<SecureString, password for accessing the pfx file>
##Example on how to set your syslog client with the certificate
for mutual authentication.
##This example script must be run from your hardware lifecycle
host or privileged access workstation.

$ErcsNodeName = "<yourPEP>"
$password = ConvertTo-SecureString -String "<your cloudAdmin
account password" -AsPlainText -Force

$cloudAdmin = "<your cloudAdmin account name>"
$CloudAdminCred = New-Object System.Management.Automation.
PSCredential ($cloudAdmin, $password)

$certPassword = $password
$certContent = Get-Content -Path C:\
cert\<yourClientCertificate>.pfx -Encoding Byte

$params = @{
    ComputerName = $ErcsNodeName
    Credential = $CloudAdminCred
    ConfigurationName = "PrivilegedEndpoint"
}

$session = New-PSSession @params

$params = @{
    Session = $session
    ArgumentList = @($certContent, $certPassword)
}
Write-Verbose "Invoking cmdlet to set syslog client
certificate..." -Verbose
Invoke-Command @params -ScriptBlock {
    param($CertContent, $CertPassword)
```

```
    Set-SyslogClient -PfxBinary $CertContent -CertPassword
$CertPassword }
```

User workloads can also be connected to SIEM tools, but in this instance, there are two options. The first is to use an external SIEM tool through the use of an agent that is deployed as part of the user workload. The second option is to make use of a SIEM tool available from the **Marketplace**. Both of these options are shown in the following diagram:

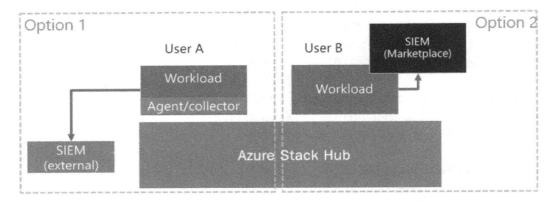

Figure 5.3 – SIEM integration for Azure Stack Hub user workloads

My final thought on security before we leave this chapter is to remember that security is a never-ending journey. With every update release that Microsoft provides, there will be security updates and Windows Defender updates.

Summary

This chapter has provided us with comprehensive coverage of all of the components within Azure Stack Hub that are used for security. We started the chapter with a look at the security infrastructure that underpins Azure Stack Hub. We learned about the security vision for Azure Stack Hub as defined by Microsoft. We then walked through the two security postures for Azure Stack Hub.

We covered the hardened by default posture, which is built from the hardware upward. We saw how to reduce the software attack surface, which included things such as ACLs, BitLocker, and certificates. We learned how to use BitLocker to encrypt the storage and also how to rotate certificates. We talked about the assume breach posture and also covered a number of Microsoft security best practices.

We then discussed the benefits of a closed system before moving on to look at Azure Stack Hub compliance with internationally recognized standards such as PCI-DSS. Finally, we took a walk through SIEM tools and how they can integrate with Azure Stack Hub.

The next chapter we are going to move on to in this book will now walk us through integrating Azure Stack Hub into the data center.

6
Considering DevOps in Azure Stack Hub

In this chapter, we are going to cover one of the use cases that are seeing an increase in the adoption of Azure Stack Hub, and that is DevOps. DevOps refers to the practice of operations and development engineers participating together in the entire service life cycle, from design, through the development process, to production support. We will explain what DevOps is from a Microsoft standpoint and how this is supported in Azure Stack Hub.

We will explain how Azure Stack Hub can be used to give you the ability to stand up a hybrid **continuous integration/continuous deployment (CI/CD)** pipeline. We will ensure that by the end of this chapter, you have a clear understanding and knowledge of DevOps with Azure Stack Hub.

We will cover the following topics in detail in this chapter:

- Introducing concepts and practices
- Reviewing DevOps on Azure Stack Hub
- Understanding the DevOps ecosystem
- Overviewing testing and metrics for DevOps

Our first section will introduce us to the concepts and practices that constitute DevOps.

Introducing concepts and practices

The best place to start with DevOps is to answer the question: *What is DevOps?* DevOps is more than just about technology—it is also about people and processes. In fact, Microsoft explains DevOps as a union of people, processes, and technology, to enable CD of value to your end users. The following diagram gives a good pictorial representation of this statement:

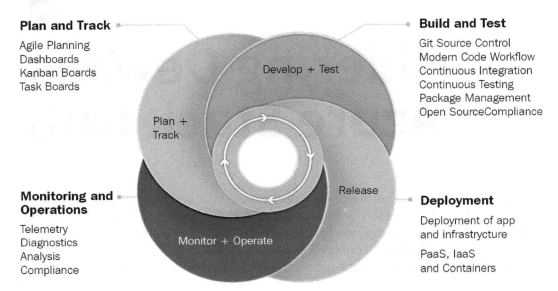

Plan and Track

Agile Planning
Dashboards
Kanban Boards
Task Boards

Build and Test

Git Source Control
Modern Code Workflow
Continuous Integration
Continuous Testing
Package Management
Open SourceCompliance

Monitoring and Operations

Telemetry
Diagnostics
Analysis
Compliance

Deployment

Deployment of app and infrastrycture

PaaS, IaaS and Containers

Figure 6.1 – What is DevOps explained

The preceding diagram demonstrates CD, starting with planning and tracking. It moves through development and testing, is released or deployed, and is then monitored as it is in operation. This process repeats as applications are updated over time. The people, processes, and technologies you see in each step are what make up DevOps.

There are seven key practices that make up the primary steps to adopting DevOps, regardless of the technology stack that is used. These seven primary steps are listed here:

- **Infrastructure as Code (IaC)**
- CI
- Automated testing
- CD
- **Release management (RM)**
- **Application performance monitoring (APM)**
- **Configuration management (CM)**

Let's take a look at each of these steps in turn, starting with IaC.

IaC

IaC is defined as a process of managing and provisioning data centers through machine-readable definition files, as opposed to physical hardware configuration or manual configuration. The infrastructure that is managed and provisioned by this process may comprise both physical machines and **virtual machines (VMs)** or environments. The provisioning of these environments may be automated, and the definition files are likely to be held in a **version control system (VCS)**. The definition files may be templated to allow the same definition files to be used for multiple environments. The next diagram shows the three pillars associated with IaC:

Templates   Automation   Versioning

Figure 6.2 – IaC pillars

This is the first step toward DevOps as it removes the complexity and human error associated with the configuration of infrastructure. It also gives us a point to start from for the next topic: CI.

CI

CI is the next step toward DevOps. CI from a development standpoint is the adoption of merging a developer's changes to a main development branch in a change control system at least daily. This then triggers a build that checks the code and offers instant feedback. The idea behind CI is to make this feedback loop as short as possible and get away from the old Waterfall approach of integrating code after months. CI is designed to work best with automated testing, which we will cover in the next topic.

Automated testing

Automated testing is generally used in conjunction with CI to offer an immediate feedback loop to a developer when they check their code into a development repository. Automated tests generally have two different functions. The first is local tests that run against the developers' code locally with no integration. Once this code is checked in it is then built, and more tests are run against this code. These tests may be more penetrative, and integrate with other systems. CI and automated testing go hand in hand with each other and lead us onto the next step in the chain, which is CD.

CD

CD takes the artifacts that have been built using CI and tested with automated tests and deploys or provisions them into an environment to allow **user acceptance testing (UAT)** to commence. It is closely related to CI and automated testing. In fact, without CI or automated testing, there is no CD. The interaction of these steps is shown in the following diagram:

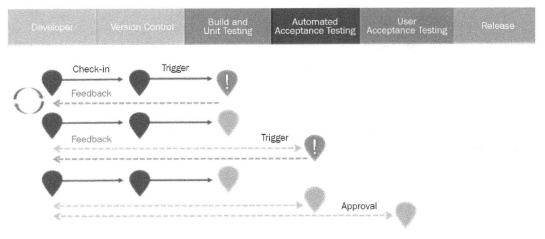

Figure 6.3 – CD steps

CD is normally used in lower-level environments where there is generally less control. As the code moves up to higher environments such as staging and production, more control is required to release the code. You will generally see CI and CD both referenced and referred to as CI/CD, which brings us to the next step: RM.

RM

RM is a process of packaging code that has been tested in lower-level environments and deploying it into a staging or production environment. This is deployed in a controlled manner, rather than auto-deployed, and careful attention is paid to the version of artifacts that are being released. This flow follows on from the CD step, as can be seen in the next diagram:

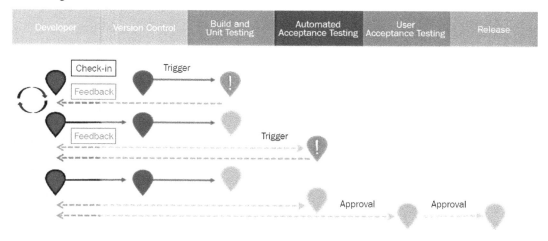

Figure 6.4 – RM flow

As you can see from this diagram, there are decision points and approvals to allow the code to move on to the next environment. This process will be repeated throughout the environments until it is released into production. Once RM has completed and the code is in production, we move on to the next step of DevOps, with APM.

APM

APM is about ensuring that once an application is published into production, it is running optimally. APM increases the detection and remediation of any errors that are found while the application is running in production and ensures that resources are being utilized effectively. It can be measured based on the uptime of the application and the number of faults found while the application is processing. This leaves us with the final step from the initial steps of DevOps, which is CM.

CM

CM brings us full circle back to IaC, where we started this topic on DevOps. There are files used to configure the physical and virtual environments that will require changes as the code passes through the different environments before reaching production. Some of these will be populated through the use of IaC, while others may be updated by operators, especially when it comes to production environments. All of these configuration files need careful management to ensure the correct configuration values are used in the right environment. This is the seventh and final step for moving to DevOps.

This is not the end of DevOps, though, but merely a beginning. These seven practices can be extended through more practices to improve the DevOps processes.

These extended practices include the following:

- Availability monitoring
- Load testing
- Auto scaling
- Feature flags
- Automated environment deprovisioning
- Self-service
- Automated recovery
- Hypothesis-driven development
- A/B testing
- Canary deployments
- Blue/green deployments

Let's take a quick look at these extended practices before we move on, starting with availability monitoring.

Availability monitoring

As with performance monitoring, availability monitoring is used to ensure that an application is configured and available in production based on uptime. Availability monitoring is used to help keep system downtime to a minimum.

Load testing

Load testing is designed to stress test the application under load. This is typically run in an environment that is identical to production, to ascertain the breaking point of an application so that adjustments can be made before deploying the application into production.

Auto scaling

Loosely related to load testing, this is a configuration to enable an application to extend its footprint based on load. This is typically used on web-based applications that spin up extra nodes when user access grows.

Feature flags

Feature flags are used to toggle features of an application on or off. This allows an organization to revoke a feature without having to do a complete rollback.

Automated environment deprovisioning

This practice normally goes hand in hand with CD and is used to ensure that an environment is stripped back to a baseline before a new release is deployed. Automating the deprovisioning of an environment ensures that the environment is in a known state and removes any human error when rolling back a previous release.

Self-service

One of the key intentions of DevOps is to give more power to users and thereby relieve pressure on the **Information Technology** (**IT**) department. The best approach for this is to implement self-service, which can allow users to deploy their own VMs, for example.

Automated recovery

Automated recovery or self-healing is a process of an application recovering from an error with no user intervention. This can be accomplished through the use of tools such as **Microsoft Operations Manager**, which can perform actions based on triggers from diagnostics logs.

Hypothesis-driven development

Hypothesis-driven development is a methodology for prototyping an application and features in production. This methodology allows developers to iteratively build, test, and rebuild their product based on user feedback.

A/B testing

A/B testing is a way of testing features within an application for various reasons such as usability, popularity, noticeability, and so on, and to look at how these factors influence the bottom line. This type of testing is normally associated with the **user interface** (**UI**) parts of an application.

Canary deployments

Canary deployments are a way of sending out a new version of an application into production to play the part of a canary, to gain insight as to how it will perform when fully integrated with other applications.

Blue/green deployments

Blue/green deployments are used for releasing applications in a predictable manner, with the goal of reducing any downtime associated with a release. Simply put, this requires having two identical environments, whereby the green environment contains the current production applications and any new releases are deployed to the identical blue environment. This allows you to quickly roll back any changes should the need arise.

This covers the DevOps practices that can and are being used by organizations, so let's turn our attention to reviewing DevOps on Azure Stack Hub.

Reviewing DevOps on Azure Stack Hub

Azure Stack Hub is a cloud-based system, and cloud-based systems ease the adoption of DevOps practices. DevOps is made easier with Azure Stack Hub, especially for the following practices:

- Self-service
- CD
- IaC

Azure Stack Hub accomplishes this with a broad ecosystem of tooling, which we will cover in detail in the next section.

Traditional on-premises workflows build and define infrastructure separately from the software, which invariably leads to multiple back and forth between the IT department defining the infrastructure and the developers writing the code that runs on the infrastructure.

This disconnect between IT departments and developers leads to multiple handoffs between them, which in turn leads to longer development cycles and potential downtime for developers while releases are handled by the IT department.

With on-premises deployments, the code is generally written for a specific platform, which leads to a lack of mobility, extendibility, and scalability.

A traditional on-premises workflow is shown in the following diagram:

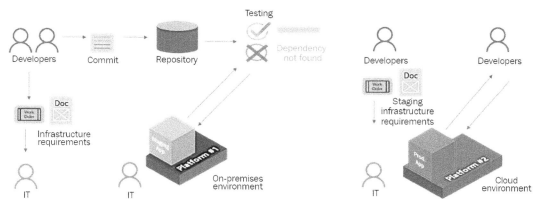

Figure 6.5 – Traditional on-premises development workflow

The difference when using DevOps in Azure Stack Hub is that the application infrastructure is defined and delivered as code, which leads to less back and forth with the IT department, resulting in fewer errors. With CD the release cycle is greatly reduced, leading to better interaction with the users. With the code able to be deployed to both Azure Stack Hub and Azure with no code changes, then mobility, extendibility, and scalability are greatly enhanced when compared to the on-premises development workflow. This reduced DevOps workflow with Azure Stack Hub is shown in the following diagram:

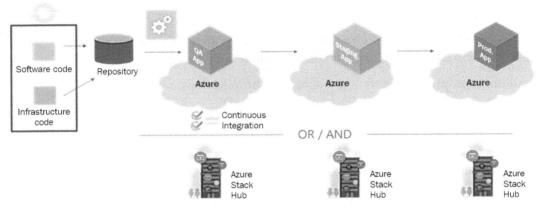

Figure 6.6 – Azure Stack Hub DevOps workflow

This DevOps workflow provides consistent application development, allowing you to build and deploy applications the same way for any Azure cloud, be that Azure Stack Hub or the public Azure cloud. You can use the same tools across Azure clouds and adopt the common DevOps practices we covered in the previous section.

To assist with the adoption of DevOps within Azure Stack Hub, Microsoft helps with the introduction of the DevOps pattern for Azure Stack Hub. This predefined pattern allows you to build, test, and deploy applications that can run on multiple clouds. This pattern includes the practices of CI and CD. As covered in the previous section, CI ensures that code is built and tested every time it is checked into the source control system. This is then improved with CD, which automates everything needed to move the application through the environments with no code changes. The only thing that changes between the environment are the configuration settings, which are environmental. This also means that the same code can be deployed to both on-premises Azure Stack Hub and Azure, with no changes or tools.

Azure Resource Manager (ARM) is used as part of this DevOps pattern, as IaC is built in and is consistent between Azure Stack Hub and Azure, as shown in the following diagram:

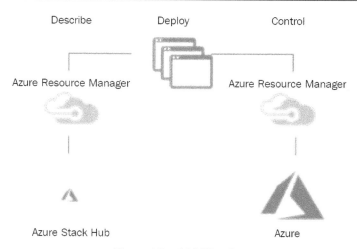

Figure 6.7 – ARM DevOps

Azure Stack Hub and Azure make the adoption of DevOps easier through the use of a consistent development environment that is built on both service and tooling consistency.

We have already seen that ARM is consistent across both Azure Stack Hub and Azure, but it is by no means the only service that is consistent across the Azure platforms. This consistency is shown in the next diagram:

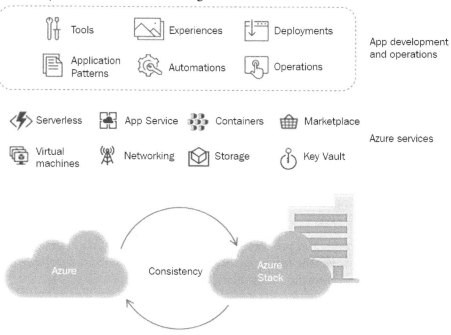

Figure 6.8 – Azure services consistency

All of this service consistency eases the adoption of DevOps, and this is further helped by the consistent tooling that is common across both Azure Stack Hub and Azure. This ensures that developers and operators do not have new tools to learn as they move between the different versions of Azure on premises and in the cloud. These consistent tools are shown in the following diagram:

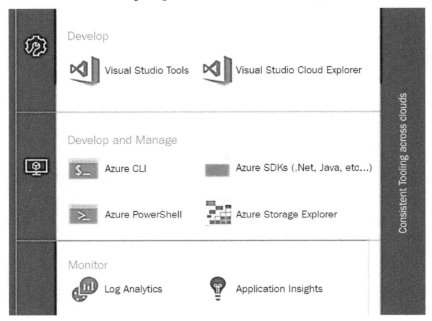

Figure 6.9 – Azure tooling consistency

Now we have reviewed DevOps on Azure Stack Hub and looked at the consistent services alongside tooling, it is time to move on to look at the DevOps ecosystem in the next section.

Understanding the DevOps ecosystem

In this section, we are going to look at the ecosystem offered by Microsoft to support DevOps with Azure Stack Hub.

Microsoft offer a myriad tools and processes, both from their own brand but also from multiple partners, which cover the full range of DevOps activities, from development through deployment, to monitoring in production.

If we begin with development from the developer's workstation, then Microsoft provides Visual Studio for writing the code and **Azure DevOps Server** for their source control, as shown in the following diagram:

Figure 6.10 – Microsoft development tools

Azure DevOps Server was previously known as **Team Foundation Server** (**TFS**) and is a set of collaborative software development tools, hosted on-premises. It integrates with an existing **integrated development environment** (**IDE**) or editors such as Visual Studio.

This environment is familiar to developers who already write applications for a cloud environment such as Azure. This is also familiar to developers who are used to working in a Windows environment. But this is not the only option for a development environment. When you look beyond the standard Microsoft tooling, then the development ecosystem expands to include applications such as GitHub, Docker, and so on, as shown in the following diagram:

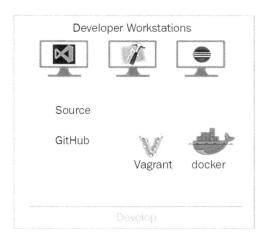

Figure 6.11 – Heterogeneous development tools

GitHub is an internet-hosted software development and source control repository.

Docker is a **platform-as-a-service (PaaS)** product that uses virtualization to deliver software in packages called containers.

Vagrant is an open source software product for building and maintaining portable virtual software development environments.

This environment is familiar for developers who are used to working in heterogeneous environments and not just Microsoft environments.

As we move through the DevOps process, then this pattern is repeated with both Microsoft and non-Microsoft options available within the Azure Stack Hub DevOps process.

The next step in the DevOps process is the build process. From a Microsoft standpoint, this continues the use of **Azure DevOps Server** and **Release Management for Visual Studio**, as shown here:

Figure 6.12 – Microsoft DevOps build tools

As with a development environment, this is acutely familiar to Microsoft developers.

For a heterogeneous build environment, then **Grunt** and **Gradle** are bought into play in place of **Azure DevOps Server**, as shown here:

Figure 6.13 – Heterogeneous build tools

Grunt is a tool used to automatically perform frequent tasks such as compilation and unit testing.

Gradle is a build automation tool for multi-language software development.

From build, we move on to the test process, which again from a Microsoft standpoint focuses on **Azure DevOps Server** and **Microsoft Test Manager**, as shown here:

Figure 6.14 – Microsoft test tools

For a heterogeneous environment, then the test tool of choice is **Jenkins**, which is shown here:

Figure 6.15 – Heterogeneous test tools

Jenkins is an open source build, testing, and deployment automation tool.

Moving on to the deployment part of the DevOps process, then from a Microsoft viewpoint, tools such as ARM are in the equation, as shown here:

Figure 6.16 – Microsoft deployment tools

A heterogeneous toolset for deployment brings in Chef and Puppet, among others, as shown here:

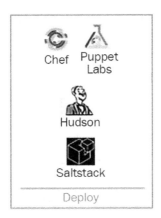

Figure 6.17 – Heterogeneous deployment tools

Chef is a configuration management tool written in Ruby that allows the writing of system configuration recipes.

Puppet is another configuration management tool that includes its own declarative language for describing system configuration.

This tooling is becoming commonplace in many environments, not just in Microsoft Azure Stack Hub.

The options for deploying developed code from an environmental standpoint include Microsoft products such as **.NET** and **SQL Server**, which are shown here:

Figure 6.18 – Microsoft environments

A heterogeneous environment includes the Microsoft options, but then also brings in Linux, Hadoop, Java, Docker, and so on, as shown here:

Figure 6.19 – Heterogeneous environments

The final step in the DevOps process for Azure Stack Hub is the monitoring step. For the Microsoft DevOps environment, this brings in Application Insights and **Operations Management Suite (OMS)**, as shown here:

Microsoft System Center
Application Insights
Operations Management Suite

Monitor and Learn

Figure 6.20 – Microsoft monitoring tools

Application Insights can only be leveraged for monitoring from Azure when the Azure Stack Hub is deployed in a connected scenario, as it can only use Application Insights from the Azure public cloud.

A heterogeneous monitoring toolset includes **Nagios**, which is shown here:

Figure 6.21 – Heterogeneous monitoring tools

All of these tools can be underpinned by processes that can be run through **Microsoft Azure DevOps Server** and **Release Manager for Visual Studio**. The list of tools that make up the DevOps ecosystem continues to grow, and I have only shown the most common ones in use in this section. The following screenshot shows the current ecosystem available in Azure Stack Hub, specifically for DevOps:

Figure 6.22 – DevOps ecosystem

As you can see from this section, the ecosystem available for Azure Stack Hub is thriving, and here, I show some other tooling that can be used:

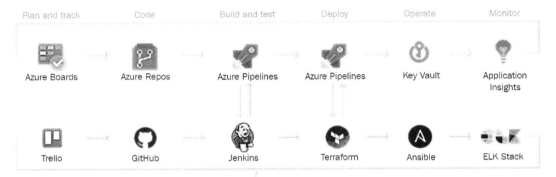

Figure 6.23 – Azure Stack Hub DevOps process with tooling

Before we move on from the DevOps ecosystem, we also need to take a look at Terraform, which is a key component for IaC.

Terraform is a tool for building, changing, and versioning cloud infrastructure safely and efficiently. Terraform codifies infrastructure in configuration files that describe the topology of relevant cloud resources. These can include a lot of the Azure Stack Hub resources we are familiar with, such as VMs, storage accounts, and networks.

From a configuration standpoint for DevOps, PowerShell and PowerShell **Desired State Configuration** (**DSC**) are also available for use with Azure Stack Hub. PowerShell DSC is a management platform that allows you manage your IT and development infrastructure with **configuration as code** (**CaC**). PowerShell DSC is a declarative platform that is used for configuration, deployment, and management of systems such as Azure Stack Hub.

One of the newer applications that have been added by Microsoft is Azure Arc. Azure Arc can also have a role to play in DevOps as it is truly hardware-agnostic, which means that developers will be able to manage their containerized applications whether running in Azure, Azure Stack Hub, or any other cloud or on-premises environment.

This completes the section on the Microsoft Azure Stack Hub DevOps ecosystem, and we will finish this chapter with an overview of testing and metrics for DevOps.

Overviewing testing and metrics for DevOps

When Azure Stack Hub is used with Azure DevOps Server during the development life cycle with DevOps, then this brings into play the functionality within Azure DevOps Server for planned manual testing, UAT, exploratory testing, and getting feedback from stakeholders.

Planned manual testing involves creating test plans and test suites that can be run by testers manually against a release of code in a test environment.

UAT takes this a step further, as real users verify that the functionality delivered by an application matches the requirements they have designated and works as designed.

Exploratory testing is testing that is generally undertaken by the development team with no set test plans and is more about exploring the functionality of the software.

Stakeholder testing is testing undertaken by users outside of the development team who will offer feedback on the functionality and usability of an application. This is normally done by a different set of users than those who conduct UAT.

These different types of manual testing and associated testers are shown in the following diagram:

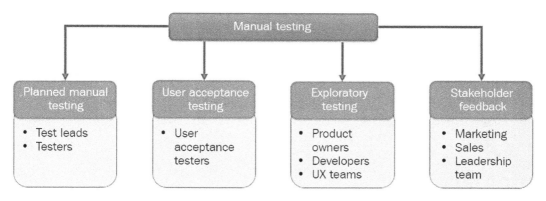

Figure 6.24 – Manual testing personas

Following on from manual testing, the next step is automated testing. Again, the use of Azure DevOps Server and Azure Test Plans makes it easy to run automated tests against test plans that have been created when looking at manual tests. Azure Test Plans in Azure DevOps Server has replaced the old Microsoft Test Manager, which was shipped with Visual Studio 2017. Azure DevOps Server provides a comprehensive test management suite that can be used both for manual tests, as shown in the previous diagram, and automated tests.

The final section we are covering in this chapter concerns metrics that can be captured as part of the DevOps process. These metrics are essentially split into two sections, as follows:

- Agility performance indicators
- Reliability performance indicators

Agility performance indicators are covered by metrics on deployment frequency and change lead time, and are designed to provide details on the team performance during the development phases. Reliability performance indicators are covered by metrics on change failure rate and mean time to detect and repair. These metrics are designed to provide details of the effectiveness of the solutions that are delivered into production.

In addition to these metrics, Azure DevOps Server also allows you to track other key metrics for the DevOps process, including the following ones:

- Failed builds
- Successful builds
- Build durations
- Completed tests

Azure DevOps can be run on-premises through the use of Azure DevOps Server or consumed as a service from Azure using Azure DevOps Service. The services available for DevOps are consistent between the versions, allowing full use of the functionality, including Azure Test Plans, Azure Pipelines, and Azure Artifacts.

This completes our walkthrough of DevOps within Azure Stack Hub and marks the end of this chapter, but before we move on to the next chapter, let's have a quick recap of what we have learned.

Summary

In this chapter, we have taken a look through DevOps and how it relates to Azure Stack Hub. We started with an understanding of key concepts and practices for DevOps. We then walked through the DevOps process in Azure Stack Hub, including IaC and CD. We then looked at the rich DevOps ecosystem provided by Azure Stack Hub, which incorporates tooling from Microsoft and a myriad partners.

We then finally looked at the DevOps testing and metrics provided for Azure Stack Hub through the use of Azure DevOps Server. This chapter has taught us everything we need to know about DevOps in Azure Stack Hub, which is a driving force in the uptake of Azure Stack Hub in many organizations.

We will now move on to look at the features in Azure Stack Hub, starting in the next chapter, where we cover ARM templates, so please join me there.

Section 3: Features

The objective of this part of the book is to cover the features of Azure Stack Hub that form the majority of the responsibility of Azure Stack administrators and operators.

The following chapters will be covered under this section:

7
Working with Resource Manager Templates

This chapter of the book is all about **Azure Resource Manager (ARM)** with **Azure Stack Hub**. We will walk through how to author ARM templates and explain how best to set up your development environment.

This chapter will give you an insight into not only the concepts for ARM, but also run through the standard JSON templates with an explanation of the different sections in the template, including resource groups and resource tags. By the end of this chapter, you should be more than familiar and confident with ARM and be able to author your own ARM templates.

This chapter will consist of the following sections:

- Understanding ARM concepts
- Authoring ARM templates
- Setting up your environment

We will begin this chapter with an understanding of ARM concepts.

Technical requirements

You can view this chapter's code in action here: `https://bit.ly/3zhreYu`

Understanding ARM concepts

ARM is an expositional, template-driven approach that serves to orchestrate deployments across various resources. ARM helps to map configuration to appropriate resource settings. It is the central endpoint to deploy models for cloud service resources and manage the life cycle and global properties of these cloud services.

ARM provides the where, what, and how through description, provision, and control. It is the management and deployment service for both Azure and Azure Stack Hub. It helps by providing a dedicated management layer that allows you to create, update, and delete resources within the Azure subscription. It enables you to use management features, such as access controls, locks, and tags, to secure and organize the resources post-deployment.

ARM provides a consistent management layer that describes the resource inventory and component relationships using tags, links, and groups. It helps to provision across regions, and across resources either in service or as a guest. It enables you to control access using **RBAC**, subscriptions, and locks. It ensures the same deployment and management experience from the portal, PowerShell, the Azure CLI, and Visual Studio.

ARM is an application life cycle container that enables you to deploy and manage your application, or resources, as you see fit.

ARM is a declarative solution for provisioning and configuration, allowing for simplified deployment of multiple instantiations of your applications via a **JSON** template.

The management and provisioning of ARM resources can be performed with any of the following:

- A REST API
- A management portal
- PowerShell
- The Azure CLI
- ARM templates

Each of these deployment mechanisms uses the ARM API and is shown in the following diagram:

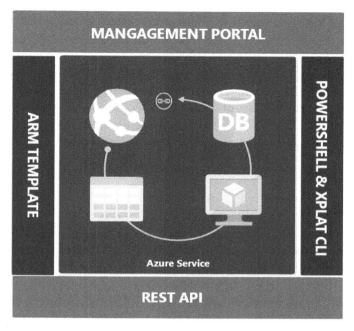

Figure 7.1 – ARM deployment mechanisms

Now that we have an understanding of what ARM is, let's turn to some of the components that make up ARM, starting with resource providers.

Resource providers

A resource provider is a service that provides the objects you can deploy and manage through ARM. As an example, to deploy an **Azure Key Vault** instance for storing keys and secrets, you would use the **Microsoft.KeyVault** resource provider.

To deploy and manage your infrastructure, you will need to know details about all of the resource providers:

- What resource types it offers
- Version numbers of the REST API operations
- Operations the resource provider supports
- A schema to use when setting the values of the resource type to create

Continuing with the resource theme, next we look at resource groups.

Resource groups

Resource groups are tightly coupled containers of multiple resources of similar or different types. Each resource can exist in one, and only one, resource group. Resource groups follow RBAC rules. When populating resource groups with resources, a decision needs to be made as to whether the resources should be placed in the same resource group or separate resource groups. The easiest way to help decide this is to think about whether they share a common life cycle and method of management. For example, in Visual Studio for developers, it is an application, while for *DevOps*, it is an environment.

The final component of the resource theme is resource tags.

Resource tags

Resource tags are loosely coupled user- or system-defined categorizations and are arbitrary boundaries. There are 15 tags available for use and are defined by the platform and user. Resource tags are name-value pairs that can be assigned to either resources or resource groups. They can be pre-created to assist with autocompletion for tag consistency. Resource tags are used for categorization or to add customer metadata. As an example, it can be used for categorizing resources based on a department or cost center, such as `tagname = "department"` and `tagvalue = "finance"`. The management portal will show views of resources organized by tags that will include monitoring and billing lenses related to those resources.

Resource tags and resource groups are the building blocks used to define applications for deployment to Azure and Azure Stack Hub.

The final component for ARM is the **REST API**.

REST API

ARM uses the REST API to manage the following resources:

- Linked resources
- Management locks
- Policy assignments
- Policy definitions
- Resources
- Resource groups

- Resource providers

- Role-based access control

- Subscriptions

- Tags

- Template deployments

- Tenants

This is shown in the following diagram:

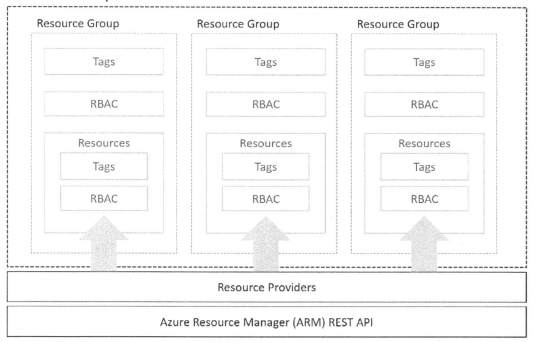

Figure 7.2 – ARM REST API

You are responsible for ensuring that any requests through the REST API are secure by using service principals, for example.

Before we move on from the ARM concepts, let's have a quick review of the components for a REST API request/response.

A REST API request/response is split into five different components:

- Request URI, which consists of `{URI-scheme}://{URI-host}/{resource-path}?{query-string}`. The request URI is part of the request message header, but most services need you to pass it separately:

  - **URI-scheme** relates to the transport protocol for transmitting the request, such as HTTP or HTTPS.

  - **URI-host** is the domain name or IP address of the server that hosts the REST API endpoint.

  - **resource-path** is used to specify the resource or resource collection.

  - **query-string** is an optional parameter that provides additional simple parameters such as API version.

- HTTP request message header fields:

  - **HTTP method** or **operation**, which tells the REST service what type of operation is being requested. This API supports `GET`, `HEAD`, `PUT`, `POST`, and `PATCH`.

  - **Optional header fields** as needed for the specified URI and method.

  As an example, an authorization header may be included that provides a bearer token containing client authorization information for the request.

- HTTP request message body, which is an optional set of fields as required by the URI and method.

- **HTTP response message header fields**:

  - **HTTP status code**, which includes 2xx success codes and 4xx and 5xx error codes

  - **Optional header fields**, as needed, to support the response, such as content-type

- HTTP response message body, which is an optional set of fields as required by the response for returning data.

Resource Manager can limit the number of read and write requests per hour to prevent an application from sending too many requests and potentially affecting other applications. If an application exceeds these limits, then the following requests from that application will be throttled. One of the fields included in the response message header fields includes the remaining number of requests for your application's scope.

An example of a `PUT` request for an ARM provider using a request header and body is shown here:

```
PUT /subscriptions/.../resourcegroups/ExampleResourceGroup?api-
version=2016-02-01  HTTP/1.1
Authorization: Bearer <bearer-token>
Content-Length: 29
Content-Type: application/json
Host: management.azure.com

{
   "location": "West US"
}
```

This covers the main components and concepts for ARM, which sets us up to move on to the next section, where we walk through the process of authoring ARM templates in detail.

Authoring ARM templates

In this chapter, we are going to walk through the process of creating ARM templates to allow us to make calls through the ARM REST API.

When authoring ARM templates that are to be used for deployment in Azure Stack Hub, then only Microsoft Azure services that are already available, or in preview, in Azure Stack Hub must be included.

Azure Stack Hub uses different service endpoint namespaces compared to Azure and, as a result, hardcoded public endpoints will cause ARM templates to fail when deployed to Azure Stack Hub. Service endpoints can be built dynamically by using reference and concatenate functions that allow the retrieval of values from the resource provider during deployment.

Azure Stack Hub resource manager functions allow you to build dynamic templates and can be used for tasks such as the following:

- Concatenating or trimming strings

- Referencing values from other resources

- Iterating on resources to deploy multiple instances

Neither **skip** nor **take** resource manager functions are available within Azure Stack Hub.

ARM templates incorporate a location attribute, which is used to place resources during deployment. This attribute in Azure would refer to a region such as West US. Locations within Azure Stack Hub are different as they are hosted in your data center. To be able to ensure that templates are transferable between Azure and Azure Stack Hub, then the resource group location should be referenced when you deploy individual resources. This can be accomplished by using [resourceGroup().Location], which ensures that all resources inherit the resource group location. By way of an example, the following code makes use of this function while deploying a storage account:

```
"resources": [
{
    "name": "[variables('storageAccountName')]",
    "type": "Microsoft.Storage/storageAccounts",
    "apiVersion": "[variables('apiVersionStorage')]",
    "location": "[resourceGroup().location]",
    "comments": This storage account is used to store the VM
disks",
    "properties": {
      "accountType": "Standard_LRS"
    }
  }
]
```

The template format has a simple structure and elements are its basis, as shown here:

```
{
    "$schema": http://schema.management.azure.com/
schemas/2015-01-01/deploymentTemplate.json#,
    "contentVersion": "",
    "apiProfile": [],
    "parameters": { },
    "variables": { },
    "resources": [ ],
    "outputs": { }
}
```

Let's walk through each of the elements from the preceding template in turn:

- `$schema`: This is a required parameter that provides the location of the JSON schema file that describes the version for the template language.

- `contentVersion`: This is a required element that provides the version of the template and ensures that the correct template is being used during deployment.

- `apiProfile`: This is a combination of resource providers and API versions that allows you access to the latest, most stable version of each resource type in a resource provider package.

- `parameters`: This is an optional element and represents values that are provided at deployment time to modify resources during deployment.

- `variables`: This is an optional element that is used as JSON fragments in the template to simplify template language expressions.

- `resources`: This is a mandatory element that defines the resource types that are deployed or updated in a resource group.

- `outputs`: This is an optional element, which is for values that are to be returned after deployment.

The syntax of this template format is **JavaScript Object Notation (JSON)** and expressions can be included. Expressions can appear any place in a JSON string value and are assessed when the template is deployed. Expressions in a JSON file must be enclosed in brackets ([and]) and they will always return another JSON value. When you need to use a literal string that begins with a bracket ([), you have to use two brackets ([[). Let's take a closer look at some of these sections of the template, starting with the parameters section.

Parameters

The syntax of the `parameters` section of the template is used to specify the values you can input when deploying resources. It enables you to modify the deployment by providing values that are bespoke for a specific environment, such as development or production. Parameters are optional, but without parameters, your template will always deploy the same resources with the same names, locations, and properties. The `parameters` section of the template is shown here:

```
"parameters": {
    "<parameterName>" : {
        "type" : "<type-of-parameter-value>",
        "defaultValue" : "<optional-default-value-of-
parameter>",
```

```
        "allowedValues" : ["<optional-array-of-allowed-
values>"]
    }
}
```

Parameters are used throughout the template to set values for the resources deployed. Only parameters that are declared in the `parameters` section can be used in other sections of the template. You cannot use a parameter value to construct another parameter value within the `parameters` section. Shown here is a working example of the `parameters` section:

```
"parameters" : {
    "siteName" : {
        "type" : "string"
    },
    "siteLocation" : {
        "type" : "string"
    },
    "hostingPlanName" : {
        "type" : "string"
    },
    "hostingPlanSku" : {
        "type" : "string",
        "allowedValues" : [
            "Free",
            "Shared",
            "Basic",
            "Standard",
            "Premium"
        ],
        "defaultValue" : "Free"
    }
}
```

From the `parameters` section of the template, we will move on to look at the `variables` section.

Variables

The variables section of the template is designed to allow you to build up values that can be used anywhere in the template. They are typically based on values provided by the parameters section. As with the parameters section, the variables section is optional, but can streamline your template by reducing complicated expressions. Shown here are the parameters and associated variables in a template:

```
"parameters" : {
    "userName" : {
        "type" : "string"
    },
    "password" : {
        "type" : "secureString"
    }
},
"variables" : {
    "connectionString" : "[concat('Name-',
parameters('username'), ';Password-', parameters('password'))]"
}
```

The following example shows the parameters and variables sections combined for creating an IaaS machine:

```
"parameters" : {
    "environmentName" : {
        "type" : "string",
        "allowedValues" : [
            "test",
            "prod"
        ]
    }
},
"variables" : {
    "environmentSettings" : {
        "test" : {
            "instanceSize" : "Small",
            "instanceCount" : 1
        },
```

```
      "prod" : {
          "instanceSize" : "Large",
          "instanceCount" : 4
      }
    },
    "currentEnvironmentSettings" :
"[variables('environmentSettings')
[parameters('environmentName')]]",
    "instanceSize" : "[variables('currentEnvironmentSettings').
instanceSize",
    "instanceCount" :
"[variables('currentEnvironmentSettings').instanceCount"
}
```

From the `variables` section, we will now move on to look at the `resources` section.

Resources

The `resources` section of the ARM template defines the resources that are to be deployed or updated. This requires an understanding of resource types to provide the correct values. The `resources` section of the template contains an array of resources to deploy, and you can also define an array of child resources for that resource. Here is an example of the complex structure for the `resources` section:

```
"resources" : [
    {
        "name" : "resourceA",
        ...
    },
    {
        "name" : "resourceB",
        ...
        "resources" : [
            {
                "name" : "firstChildResourceB",
                ...
            },
            {
                "name" : "secondChildResourceB",
```

```
                ..
            }
        ]
    },
    {
        "name" : "resourceC",
        ...
    }
]
```

The final section of the ARM template we will look at is the outputs section.

Outputs

The outputs section is used to specify values that are returned from the deployment. For example, you may want to return the URI to allow access to a newly deployed resource. This syntax is shown here:

```
"outputs" : {
    "<outputName>" : {
        "type" : "<type-of-output-value>",
        "value" : "<output-value-expression>",
    }
}
```

The example shown here shows a value being returned in the outputs section relating to the site URI:

```
"outputs" : {
    "siteUri" : {
        "type" : "string",
        "value" : "[concat('http://',
reference(resourceId('Microsoft.Web/sites',
parameters('siteName'))).hostNames[0])]"
    }
}
```

Before we move on, let's see a real-world example of a JSON file for creating a storage account:

```
{
    "$schema": "https://schema.management.azure.com/
schemas/2015-01-01/deploymentTemplate.json#",
    "contentVersion": "1.0.0.0",
    "apiProfile": "2018-03-01-hybrid",
    "parameters": {
      "storageAccountType": {
        "type": "string",
        "allowedValues": [
          "Standard_LRS"
        ],
        "defaultValue": "Standard_LRS",
        "metadata": {
          "description": "Type of redundancy for your storage
account"
        }
      }
    },
    "variables": {
      "storageAccountName":
"[concat(uniqueString(resourceGroup().id),'storage')]"
    },
    "resources": [
      {
        "type": "Microsoft.Storage/storageAccounts",
        "name": "[variables('storageAccountName')]",
        "location": "[resourceGroup().location]",
        "sku": {
          "name": "[parameters('storageAccountType')]"
        },
        "kind": "Storage"
      }
    ],
    "outputs": {}
  }
```

This completes our review of the sections of the ARM template, and we will now move on to take a look at the ARM schemas.

Schemas

JSON files use a **schema**, which is referenced at the top of each file. The schema defines what elements are allowed, the types and formats of fields, the possible values of enumerated values, and so on.

There are three main ways to explore the schema and properties available:

- Published schema files
- Example or sample templates
- JSON responses from the REST APIs

From schemas, we switch to look at expressions and functions.

Expressions and functions

Expressions and functions can be used to extend the JSON available in the template. They enable you to create values that are not strict literal values.

Additionally, the following applies to expressions:

- They are enclosed within brackets ([and]).
- They are evaluated when the template is deployed.
- They can appear anywhere in the JSON string value and always return another JSON value.

Typically, you use expressions with functions to perform operations for modifying the deployment, as shown in the following example:

```
"variables" : {
    "location" : "[resourceGroup().location]",
    "usernameAndPassword" : "[concat('parameters('username'),
':', parameters('password'))]",
    "authorizationHeader" : "[concat('Basic ',
base64(variables('usernameAndPassword')))]"
}
```

You can use expressions with functions to perform operations for modifying the deployment. Function calls are defined using the `functionName(arg1,arg2,arg3)` syntax. You then need to reference properties by using the dot and [index] operators.

The previous example shows how to use several functions when building up values.

The next concept with the ARM templates we will look at is linking.

Linking ARM templates

From within one ARM template, it is possible to link to another template, which allows you to break down your deployment into a set of defined, targeted, purpose-specific templates.

You can pass parameters from the calling template to a linked template, and those parameters can map to parameters or variables exposed by the initiating calling template.

The linked template is also able to pass an output variable back to the initial calling template, facilitating a bi-directional data exchange between templates. To document and provide query capability over the relationships between resources, you should use resource linking.

To create a link between two templates, perform the following steps:

- Create a deployment resource within the calling template and point it at the linked template.

- Update the `templateLink` property and set it to the URI of the linked template.

- Optionally provide parameter values for the linked template, either by specifying the values directly in your template, or by linking to a parameter file.

This example uses the `parameters` property to specify a parameter value directly:

```
"resources" : [
    {
        "apiVersion" : "2015-01-01",
        "name" : "nestedTemplate",
        "type" : "Microsoft.Resources/deployments",
        "properties" : {
            "mode" : "incremental",
            "templateLink" : {
                "uri" : https://www.contoso.com/AzureTemplates/
newStorageAccount.json,
                "contentVersion" : "1.0.0.0"
```

```
            },
            "parameters" : {
                "StorageAccountName" : {"value" :
    "[parameters('StorageAccountName')]"}
            }
        }
    }
]
```

The **Resource Manager** service cannot access local files or local network files for the linked template. You are only allowed to provide a URI value that includes either http or https. This URL must be accessible from Azure Stack Hub, especially in the disconnected scenario. This URL could also point to a blob in an Azure Stack Hub storage account.

The following example uses the parametersLink property to link to a parameters file. The URI value for the linked parameter file cannot be a local file and must include either http or https:

```
"resources" : [
{
    "apiVersion" : "2015-01-01",
    "name" : "nestedTemplate",
    "type" : "Microsoft.Resources/deployments",
    "properties" : {
        "mode" : "incremental",
        "templateLink" : {
            "uri" : https://www.contoso.com/AzureTemplates/
newStorageAccount.json,
            "contentVersion" : "1.0.0.0"
        },
        "parametersLink" : {
            "uri" : https://www.contose.com/AzureTemplates/
parameters.json,
            "contentVersion" : "1.0.0.0"
        }
    }
}
]
```

ARM templates allow you to define dependencies between resources through the use of the dependsOn and resources properties. The dependency between resources is evaluated and resources are deployed in their dependent order. For example, if you have a web application that depends on an underlying database, then you could structure the dependency within ARM to ensure that the database is deployed first, followed by the web application.

The dependsOn property gives the ability to define dependencies on a resource as per the preceding database example or to define the dependence on multiple nodes in the cluster being installed before deploying a virtual machine with the cluster management tool. The dependsOn property is not designed to be used for documenting how resources are interconnected. The dependsOn property takes an array of related resource names.

The resources property allows you to specify child resources that are related to the resource that is being defined. Child resources can only be up to five levels deep. If the child resources need to be deployed after the parent resource, then the dependsOn property should be used as discussed. Each parent resource only allows certain resource types to be defined as children. Allowed resource types are defined in the template schema of the parent resource. The name of the child resource type includes the name of the parent resource type, such as Microsoft.Web/sites/config and Microsoft.Web/sites/extensions, where config and extensions are both children of Microsft.Web/sites.

To help with the creation of ARM templates, there are some quick-start templates available on GitHub, which gives you a great starting point. These can be found at the following URL, https://github.com/Azure/AzureStack-QuickStart-Templates, and all the templates found here are deployable to Azure Stack Hub.

Once you have created your ARM template and before you deploy it into the Azure Stack Hub environment, it is always recommended to verify that they are ready for deployment by using the template validation tool, which is available from the Azure Stack Hub tools' GitHub repository, which is located here: https://github.com/Azure/AzureStack-Tools.

From a workstation connected to Azure Stack Hub via a VPN, open a PowerShell **Integrated Scripting Environment (ISE)** window and import the CloudCapabilities module using the following cmdlet:

```
import-Module .\CloudCapabilities\Az.CloudCapabilities.psm1
```

You then create a cloud capabilities JSON file using the following command:

```
get-AzCloudCapability -Location <your-location> -Verbose
```

Then, to test the template, import the template validator and run it against your template using the following in the PowerShell window:

```
import-Module .\Az.TemplateValidator.psm1

Test-AzTemplate -TemplatePath <path to template.json or
template folder> -CapabilitiesPath <path to cloudcapabilities.
json> -Verbose
```

Validation errors and warnings for the template are displayed in the *PowerShell* console and written to an **HTML** file in the source directory. The following screenshot shows an example of a validation report:

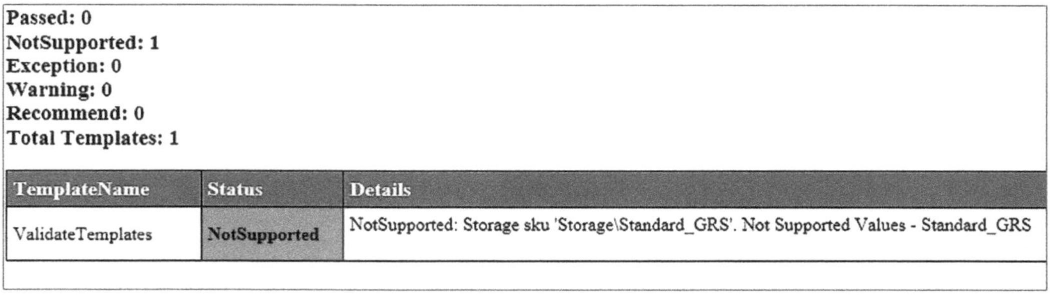

Passed: 0
NotSupported: 1
Exception: 0
Warning: 0
Recommend: 0
Total Templates: 1

TemplateName	Status	Details
ValidateTemplates	NotSupported	NotSupported: Storage sku 'Storage\Standard_GRS'. Not Supported Values - Standard_GRS

Figure 7.3 – Template validation HTML report

Now that we understand the schemas, elements, and properties of the ARM templates, we can look at how you would set up your ARM template authoring environment.

Setting up your environment

In this section of the chapter, we will look at the development environments for authoring ARM templates and consider preparing for developing ARM templates.

For developers running Windows OSes, the standard *Visual Studio* and free *Visual Studio Code* can be used to author ARM templates. In fact, any text editor can be used to author ARM templates as the underlying structure is JSON format, as we described in the previous section.

To be able to author ARM templates within Visual Studio, a little preparation is required. To be able to author ARM templates, we need to download and install the correct tools. This is easily accomplished within Visual Studio by performing the following steps:

- Create a new project.
- Select **Cloud**.
- Select **Get Microsoft Azure SDK for .NET**.
- Click **OK** and then download Microsoft Azure SDK for .NET.

Once this has been successfully downloaded and installed in Visual Studio, then we are ready to begin the authoring of ARM templates from within Visual Studio.

In Visual Studio, chose **File | New Project**, and then select **C#** or **Visual Basic**. Then choose **Cloud**, followed by the **Azure Resource Group** project as per the following screenshot:

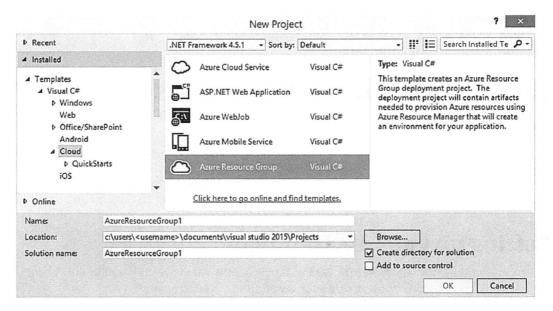

Figure 7.4 – Azure Resource Group Visual Studio project

Choose the template that you want to deploy to ARM. There are many different options based on the type of project you wish to deploy; for example, the **Web App + SQL** template, as shown in the following screenshot:

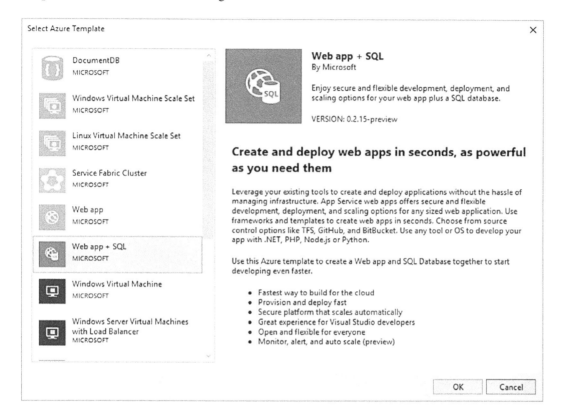

Figure 7.5 – Azure Template selection in Visual Studio

The template you select is only a starting point and you can add and delete resources to fulfill your particular scenario.

Visual Studio then creates a Resource Group deployment project for both the Web App and SQL database, as shown here:

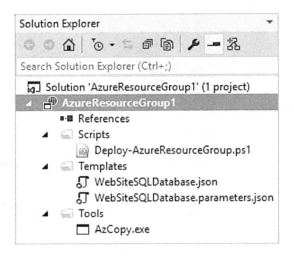

Figure 7.6 – Visual Studio Resource Group deployment project

In this example, you can see the following files:

- `Deploy-AzureResourceGroup.ps1`: This is the PowerShell script that is used to deploy to ARM via a series of embedded PowerShell cmdlets. This PowerShell script is used by Visual Studio to deploy the template, so any changes to this PowerShell script will affect deployment in Visual Studio.

- `WebSiteSQLDatabase.json`: This Resource Manager template specifies the infrastructure that will be deployed and the parameters that can be provided during deployment. It also defines the dependencies between the resources, to ensure that they are deployed in the correct order.

You can customize a deployment project by modifying the JSON template. The Visual Studio editor provides tools to assist you with editing the resource manager template. The **JSON Outline** window makes it easy to see the elements defined in your template, as shown in the following screenshot:

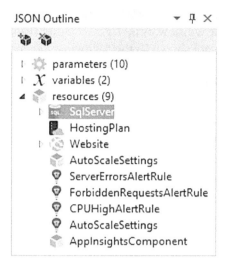

Figure 7.7 – JSON Outline window in Visual Studio

Selecting any of the elements in the outline takes you to that part of the template and highlights the corresponding JSON, as in the following screenshot, where **Website** has been selected:

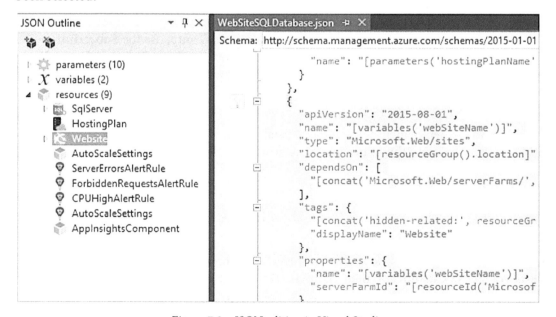

Figure 7.8 – JSON editing in Visual Studio

You can add a new resource to your template by either selecting the **Add Resource** button at the top of the **JSON Outline** window or by right-clicking **resources** and selecting **Add New Resource**, both of which are shown in the following screenshot:

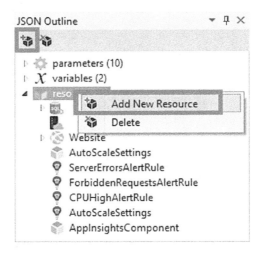

Figure 7.9 – Adding a new resource to a project in Visual Studio

It is possible to deploy the template to Azure Stack Hub directly from Visual Studio. This can be done simply by right-clicking the name of the project and selecting **Deploy**. Select **New Deployment** and then, in **Deploy to Resource Group**, use the **Subscription** dropdown to select **Microsoft Azure Stack Hub subscription**. From **Resource Group list**, choose an existing resource group or create a new one. From **Resource Group Location**, choose
a location and then select **Deploy**.

In addition to the deployment from within Visual Studio, it is also possible to deploy the ARM templates by using the **Azure portal**, **Azure PowerShell**, and the **Azure CLI**.

Deploying with the portal

It is possible to use the Azure Stack Hub *user portal* to deploy ARM templates to Azure Stack Hub.

First, you need to sign in to the Azure Stack Hub user portal and select **+ Create a resource**. Select **Custom** and then **Template Deployment**.

You can either select **Type** to filter to choose a GitHub **QuickStart** template or choose **Build your own template** in the editor.

The quickstart templates can be found at the following URL, `https://github.com/Azure/AzureStack-QuickStart-Templates`, and these are written by members of the Azure Stack Hub community and not by Microsoft.

If you choose **Build your own template** in the editor, then this will allow you to paste your JSON template code into the code window. When you are ready with your template, you can click **Save**.

Select the subscription for your Azure Stack Hub and then select or create a new resource group. Once all the settings have been selected, click **OK** followed by **Review + Create**.

Finally, select **Create**, which will deploy the template to Azure Stack Hub.

Deploying with PowerShell

It is also possible to use PowerShell to deploy ARM templates to the Azure Stack Hub by performing the following steps:

1. Save the `101-simple-windows-vm` template from the quickstart templates' GitHub repository in the previous section to a local location.

2. Run a PowerShell window as **Administrator** and use the following sample script to deploy this quickstart template:

```
$myNum = "001"
$RGName = "myRG$myNum"
$myLocation = "yourregion"
New-AzResourceGroup -Name $RGName -Location $myLocation
New-AzResourceGroupDeployment -Name myDeploymentName
-ResourceGroupName $RGName -TemplateURI <path>\
AzureStack-QuickStart-Templates\101-vm-windows-create\
azuredeploy.json -AdminUsername <username> -AdminPassword
("<password>" | Convert.ToSecureString -AsPlainText
-Force)
```

This will deploy the virtual machine based on the template and should be visible in the Azure Stack Hub portal.

Deploying with the CLI

The Azure **command-line interface** (**CLI**) can also be used to deploy ARM templates in Azure Stack Hub.

The latest version of the Azure CLI is available for download for installation in Windows from `https://aka.ms/installazurecliwindows`.

Use the following command to log in to Azure Stack Hub from the Azure CLI in a terminal window:

```
az cloud register '
    -n <environmentname> '
    --endpoint-resource-manager "https://
management.<region>.<fqdn>" '
    --suffix-storage-endpoint "<fqdn>" '
    --suffix-keyvault-dns ".vault.<fqdn>" '
    --endpoint-active-directory-graph-resource-id https://
graph.windows.net/
Az cloud set -n <environmentname>
Az cloud update -profile 2019-03-01-hybrid
az login - tenant <AAD tenant name> --service-principal -u
<Application ID of the service principal> -p <key generated for
the service principal>
az group create -name testDeploy -location local
az group deployment create -resource-group testDeploy -
template-file ./azuredeploy.json -paramters ./azuredeploy.
parameters.json
```

This example command deploys the template to the `testDeploy` resource group in your Azure Stack Hub instance. From here, let's now look at the different API version profiles.

API version profiles

API profiles specify the *Azure resource provider* and the *API version* for Azure REST endpoints. This allows you to work with Azure resource providers without having to know the exact version of each resource provider API that is compatible with Azure Stack Hub. You just need to align your application to a profile and the SDK reverts to the correct API version.

API profiles are used to represent a set of Azure resource providers with their associated API versions. They have been created to allow you to create templates across multiple Azure clouds, including Azure Stack Hub. This provides a stable interface for interacting with the Azure clouds and are released four times a year. Not all API profiles are compatible with Azure Stack Hub, so the profile to look for is the YYYY-MM-DD-Hybrid profile, which is normally released in March and September every year. This profile is designed to work against both Azure and Azure Stack Hub. The API versions listed in this profile will be the same as the ones that are listed in Azure Stack Hub. This is the profile that should be used to develop templates against Azure Stack Hub.

The API Hybrid profile includes the following resource provider connections:

- `Microsoft.Compute`
- `Microsoft.Network`
- `Microsoft.Storage` (data plane)
- `Microsoft.Storage` (control plane)
- `Microsoft.Web`
- `Microsoft.KeyVault`
- `Microsoft.Resources` (ARM itself)
- `Microsoft.Authorization` (policy operations)
- `Microsoft.Insights`

We can now take a look at each of these in turn.

Microsoft.Authorization

This lets you manage the actions whereby users in your organization can take on resources within Azure Stack Hub. With this, you can define roles, assign roles to users and groups, and retrieve information about permissions. This lets you interact with the following resource types:

- Locks
- Operations
- Permissions
- Policy assignments
- Policy definitions

- Provider operations
- Role assignments
- Role definitions

From authorization, we move on to compute.

Microsoft.Compute

This gives you programmatic access to virtual machines and their associated resources. It allows you to interact with the following resource types:

- Availability sets
- Locations
- Locations/operations
- Locations/publishers
- Locations/usages
- Locations/vmSizes
- Operations
- Virtual machines
- Virtual machines/extensions
- Virtual machine scale sets
- Virtual machine scale sets/extensions
- Virtual machine scale sets/network interfaces
- Virtual machine scale sets/virtual machines
- Virtual machine scale sets/virtual machines/network interfaces

From compute, we next look at insights.

Microsoft.Insights

This lets you interact with event and diagnostics information, including the following resource types:

- Operations
- Event types

- Event categories
- Metric definitions
- Metrics
- Diagnostic settings
- Diagnostic setting categories

After insights, we move on to look at Key Vault.

Microsoft.KeyVault

This allows you to manage your Key Vault instance as well as the associated keys, secrets, and certificates. It gives access to the following resource types:

- Operations
- Vaults
- Vaults/access policies
- Vaults/secrets

From managing Key Vault, we next look at network.

Microsoft.Network

This is a representation of the available network cloud operations' lists and is used to interact with the following resource types:

- Connections
- DNS zones
- Load balancers
- Local network gateway
- Locations
- Location/operation results
- Locations/operations
- Locations/usages
- Network interfaces
- Network security groups

- Operations

- Public IP address

- Route tables

- Virtual network gateway

- Virtual networks

From looking at networks, we move on to looking at resources.

Microsoft.Resources

ARM allows you to deploy and manage the infrastructure for your Azure Stack Hub solution. This lets you combine related resources in resource groups and deploy your resources with JSON templates. This gives you access to the following resource types:

- Deployments

- Deployments/operations

- Links

- Locations

- Operations

- Providers

- Resource groups

- Resources

- Subscriptions

- Subscriptions/locations

- Subscriptions/operation results

- Subscriptions/providers

- Subscriptions/resource groups

- Subscriptions/resource groups/resources

- Subscriptions/resources

- Subscriptions/tag names

- Subscriptions/tag names/tag values

After resources, we can now look at web.

Microsoft.Web

The web resource provider allows you to manage your sites, certificates, and server farms programmatically and provides access to the following resource types:

- Certificates
- Operations
- Checknameavailability
- Metadata
- Sites
- Sites/domainOwnershipidentifiers
- Sites/extensions
- Sites/hostNameBindings
- Sites/instances
- Sites/instances/extensions
- Sites/slots
- Sites/slots/hostNameBindings
- Sites/slots/instances
- Sites/slots/instances/extensions
- Server farms
- Server farms/metricDefinitions
- Server farms/metrics
- Server farms/usages
- Available stacks
- Deployment locations
- Georegions
- List sites assigned to host name
- Publishing users
- Recommendations
- Source controls
- Validate

After web, we can now move on to look at `Microsoft.Storage`.

Microsoft.Storage

The storage resource provider allows you to manage your storage account and keys programmatically and provides access to the following resource types:

- CheckNameAvailability
- Locations
- Locations/quotas
- Operations
- Storage accounts
- Usages

Some projects provide samples for using the API profiles, such as managing resource groups in GitHub, available from the following location: `https://github.com/Azure-Samples/hybrid-resources-dotnet-manage-resource-group`.

This completes the chapter on ARM templates, but before we move on, let's have a quick recap of what we have learned in this chapter.

Summary

In this chapter, we were introduced to ARM templates. We started with a review of some of the key concepts that help us to understand ARM templates, such as schemas and parameters. From there, we dived deep into the construction of ARM templates before rounding out the chapter by covering the development and deployment environment. This chapter should make it possible for you to now be able to set up your development environment and be comfortable creating your own ARM templates.

Please join me in *Chapter 8*, *Working with Offers, Plans, and Quotas*, where we will be covering offers, plans, and quotas in Azure Stack Hub.

8
Working with Offers, Plans, and Quotas

Following on from our last chapter, which talked about authoring **ARM** templates, we'll now move on to talk about everything we need to know to be able to offer resources and services from Microsoft Azure Stack Hub. We will explain each of the components needed to create an offer on Azure Stack Hub including resources, quotas, and plans. We will provide guidance on the steps required to create a new offer and also touch on delegation and automation. This teaches us everything we need to know to be able to offer services to tenants including the subscription process and how it relates back to the public Azure cloud.

In this chapter, we will be covering the following main topics:

- Planning service offerings
- Creating a plan
- Creating an offer
- Subscribing to an offer
- Delegating offers
- Managing plans and offers

We'll begin with a look at how we need to plan for service offerings.

Technical requirements

You can view this chapter's code in action here: `https://bit.ly/3sO7XLU`

Planning service offerings

In this first section, we will begin by looking at service offerings in general before we delve into greater detail later on in the chapter when we look at what it takes to create an offering within Azure Stack Hub. Before that, let's think about how users get a subscription for Azure. Users sign up for Azure, which in turn subscribes them to an offer such as pay as you go, an enterprise agreement, MSDN, a trial, and so on. These are offers that are provided by Microsoft for users to subscribe to when they create their Azure subscription. With Azure Stack Hub, you as the operator can then take on the role of Microsoft and must define the offers you wish to make available from your platform.

In Azure Stack Hub, you are the provider, which means that you define quotas, plans, and offers. You control the quotas specific to each offer and you define your business operating model.

You as the Azure Stack Hub operator can configure and deliver services by using offers, plans, and subscriptions. Offers contain at least one plan, and each plan will include one or more services, each of which is configured with quotas. By defining plans and combining them into different offers, users are able to subscribe to your offers and deploy resources. This allows you to control which services and resources your users can access. It also allows you to control the amount of resources they can consume and from where.

In simple terms, you deliver a service through Azure Stack Hub by creating a plan that includes one or more services using foundational services such as compute, storage, and networking, and assign quotas to limit the resources that can be consumed as part of the plan. You then create an offer that includes one or more plans. Once this offer is created, then users will be able to subscribe to this offer.

Let's walk through this process in a little more detail.

The starting point for the planning is the resource and services, as shown in this first diagram:

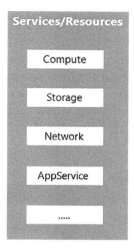

Figure 8.1 – Azure Stack Hub resources

From here, we need to define a quota against the resource or service. A quota is used to determine the upper limit of the resources the subscriber can consume. You can create multiple levels of quotas for the same resource. This is shown in the next diagram:

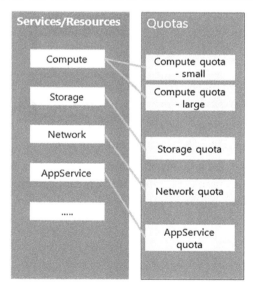

Figure 8.2 – Azure Stack Hub resource quotas

It can take several hours for new quotas to be available within the user portal or before a changed quota is enforced.

From quotas, we can create a plan, which is used to group several quotas together as a logical unit. As an example, think about a mobile phone operator. They tie together quotas for text messages, calls, and data into a plan. The same idea is used in Azure Stack Hub. You can think of plans as the levels of service capacity that you might offer users. The quotas are the granular controls, and plans tie them together into more general groupings as we show in the next diagram:

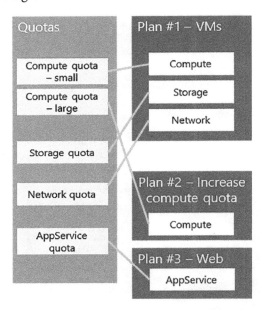

Figure 8.3 – Azure Stack Hub plans

A base plan is the default plan for a given offer and is included by default when the user subscribes to an offer. Add-on plans are optional plans that can be bolted onto an offer and are not included by default when a user subscribes to the offer.

An offer is then used to make one or more plans available to users. An offer must have at least one base plan. An offer can have a base plan and zero or more add-on plans associated with it. This means you can include multiple plans in one offer. Offers are what users see when they sign up for Azure or Azure Stack Hub and an example is shown in the next diagram:

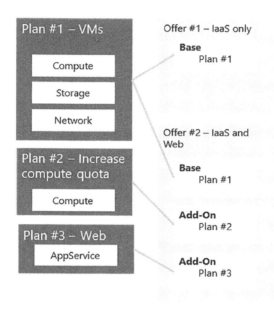

Figure 8.4 – Azure Stack Hub Offers

Offers will have one of three states in Azure Stack Hub:

- **Public**: Visible to all users
- **Private**: Visible only to cloud administrators
- **Decommissioned**: Not able to be subscribed

When thinking about offers and planning the services you wish to offer, then it is worth thinking about the following points:

- **Trial offers**: Trial offers can be used to attract new users, who will then be able to upgrade to additional services.

- **Capacity planning**: You need to carefully plan your quotas to ensure that users do not grab a large amount of resources to the detriment of other users.

- **Delegation**: It is possible to delegate the ability to create offers to users within the environment, for example, an organization granting departments the ability to create their own offers.

Once an offer is made public and is visible to Azure Stack Hub users, then they are able to subscribe to the offer and start to consume the associated resources as shown in the next diagram:

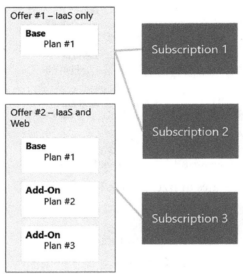

Figure 8.5 – Azure Stack Hub subscription

A subscription is how users sign up to offers within Azure Stack Hub. When a user or organization first starts to use Azure Stack Hub resources, they must choose an offer and subscribe to that offer. A customer-centric view of this is shown in the next diagram:

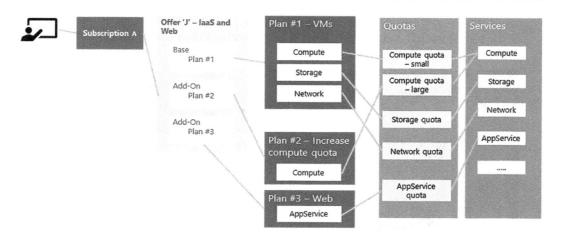

Figure 8.6 – Azure Stack Hub customer subscription

Users get access to existing subscriptions and create new subscriptions by signing in to Azure Stack Hub. A subscription represents an association with a single offer. If a user requires access to more than one offer, then they must create another subscription. As an Azure Stack Hub operator, you are able to see information about tenant subscriptions, but you are not able to access the contents of the subscription unless you have been explicitly added through RBAC by a tenant administrator of that subscription. This allows the separation of power and responsibilities between the Azure Stack Hub operator and the tenant spaces.

Now we understand what goes into planning the offering of services from Azure Stack Hub, let's walk through the process of creating these offers, starting with creating a plan.

Creating a plan

As we have seen in the previous section, plans are logical groupings of one or more services and their associated quotas. Let's walk through the process of creating a plan that includes compute, network, and storage resource providers. This sample plan will give subscribers the ability to provision virtual machines:

1. The starting point for creating a plan is to sign in to the Azure Stack Hub administrator portal, which is found at the following URL: `https://adminportal.local.azurestack.external`.

2. Once signed into the administrator portal, click the **+ Create a resource** button at the top of the left-hand pane. Then select **Offers + Plans**, and finally select **Plan**. This is shown in the following screenshot:

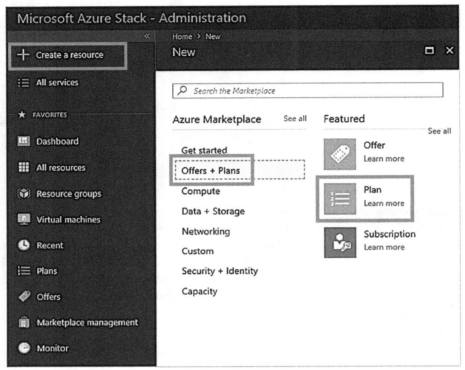

Figure 8.7 – Azure Stack Hub administrator portal – create a plan

This will then bring up a tabbed user interface, which will enable you to specify the plan name, add services, and define quotas for each of the selected services.

3. The **Basics** tab in this **New plan** window allows you to specify a display name and a resource name. The display name is a friendly name for the plan, which will be visible to Azure Stack Hub operators. The other field on this tab is **Resource group**, which will be the container for the plan. This can either be set by selecting an existing resource group or creating a new one. It is advised and is best practice to have a naming convention for all of the components, including quotas, plans, and offers. When creating these different items, it also makes sense to store them all in the same dedicated resource group. The **New plan** window with the **Basics** tab is shown in the following screenshot:

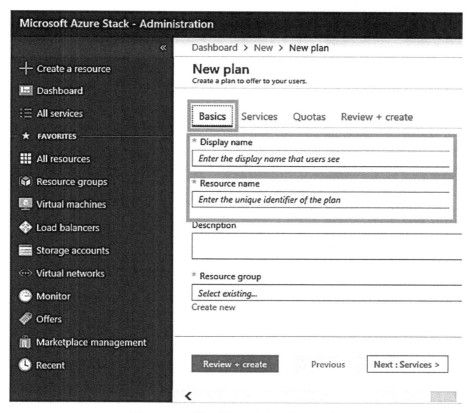

Figure 8.8 – New Plan window Basics tab

4. The **Next: Services** button will move you onto the next tab, **Services**, or alternatively, you can click the **Services** tab at the top of the window. This tab allows you to select the services to be offered as part of this plan. In the example in the next screenshot, we have selected **Microsoft.Compute**, **Microsoft.Storage**, and **Microsoft.Network**:

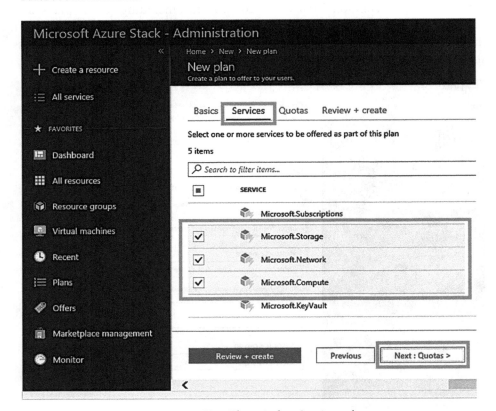

Figure 8.9 – New Plan window Services tab

5. Clicking **Next: Quotas** will take us onto the next tab, which is **Quotas**. This tab displays the services that were selected on the previous tab with a drop-down box next to each service. This drop-down box allows you to select a default quota for the given service or you can create a custom quota against the service by selecting **Create New** next to the service. The **Quotas** tab is shown in the following screenshot:

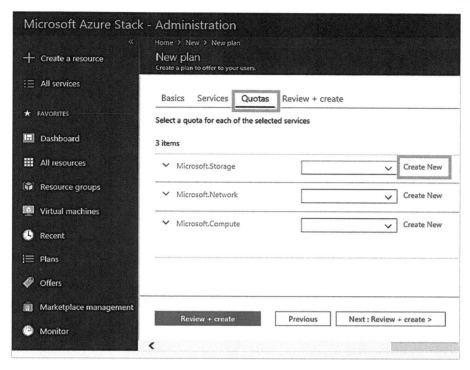

Figure 8.10 – New Plan window Quotas tab

If you click **Create New** against the service quota, then another window is displayed, which allows you to specify a name for the quota and specify the quota values based on the service that was selected. For example, the following is the **Create quota** window for the **Microsoft.Storage** service:

Figure 8.11 – New quota for Microsoft.Storage

6. Once all quotas have either been selected or created for the associated services, clicking on the **Next: Review + create** button will move onto the final tab.

7. This final tab lets you review all the values you have entered for the plan. Once you are happy with the plan, you can click **Create**, which will then create the plan.

Once the plan is created, it will be visible by selecting **All Services** on the left-hand side panel and then selecting Plans. Depending on how many plans are defined within this Azure Stack Hub instance, you may need to search for the plan by name to locate it.

This only creates the plan and plans are only visible to Azure Stack Hub operators. To make them available to users, we need to add them to an offer definition and we will walk through this process in the next section.

Creating an offer

Offers are groups of one or more plans that operators define and present to users, which those users can buy or subscribe to. Following on from the previous section on plans, we will now walk through creating an offer that makes use of the plan we worked through:

1. As with plans, the starting point is the Azure Stack Hub administrator portal, which is available at the following URL: `https://adminportal.local.azurestack.external`.

2. Select the **+ Create a resource** button in the top-left corner of the screen. Select **Offers + Plans** and then **Offer**, as shown in the following screenshot:

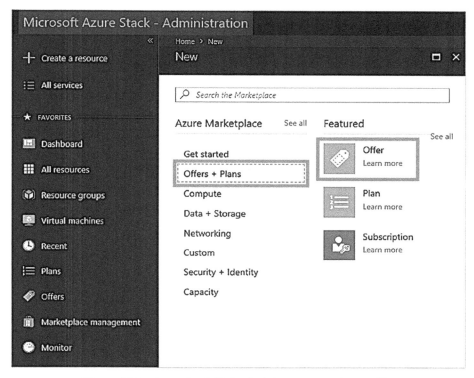

Figure 8.12 – Azure Stack Hub administrator portal new offer

3. This will then bring up a tabbed user window that allows you to define the offer name. It is also possible to add existing plans or create new base plans and add-on plans. The first tab that is shown is the **Basics** tab, where you can enter the display name and resource name. This tab is also used to select an existing resource group or create a new resource group to locate the offer. The display name is the name that users will see in the portal to subscribe to the offer to so it should be descriptive of what is included in the offer. Only administrators can see the resource name and this is the name that administrators will use to work with the offer as an Azure Resource Manager resource. At the bottom of the page is a selection for setting the offer as private or public. By default, all new offers are set to private but this can be changed at any time.

The **Basics** tab is shown in the following screenshot:

Dashboard > New > **Create a new offer**

Create a new offer
Create a new offer for your users

| **Basics** | Base plans | Add-on plans | Review + create |

* Display name ❶

Enter the display name that users see

* Resource name

Enter the unique identifier of the offer

Description

* Resource group

Select existing...
Create new

Make this offer public?

| Yes | No |

| Review + create | Previous | Next : Base plans > |

‹

Figure 8.13 – Azure Stack Hub new offer Basics tab

4. Select the **Next: Base plans** button to move to the next tab where you can select the plan(s) to be included in the offer. Within this tab, you can select existing plans or create a new base plan, as shown in the following screenshot:

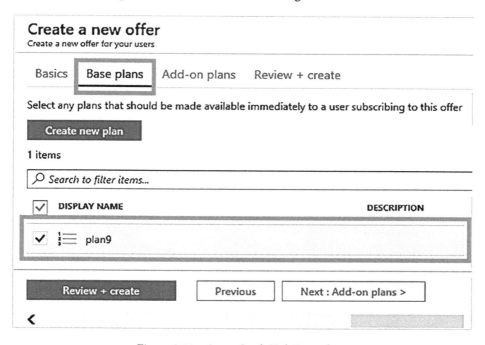

Figure 8.14 – Azure Stack Hub Base plans

5. You can click the **Next: Add-on plans** button then. This gives you the opportunity to create an add-on plan, but this is an optional step. If no add-on plans are needed, then click on the **Review + create** tab to ensure the values are correct and then click **Create** to create the offer as shown in the following screenshot:

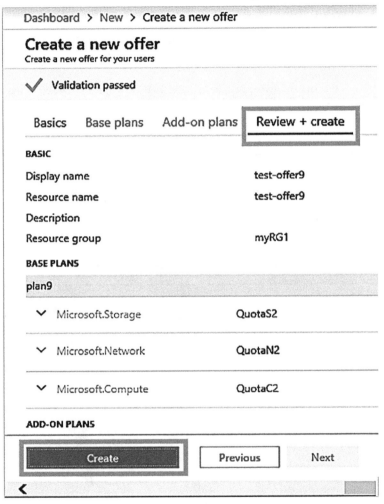

Figure 8.15 – Azure Stack Hub – create offer

Once an offer has been created, you can change the state of the offer from private to public. The offer must be public for users to be able to subscribe to the offer. As mentioned earlier in this chapter, an offer can be private, public, or decommissioned. There are two options for changing the state of the offer from the **All Resources** view. Select the name of the offer and on the overview screen, select **Change state,** as shown in the following screenshot:

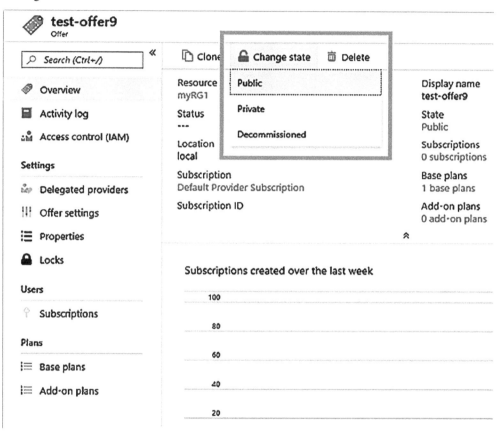

Figure 8.16 – Azure Stack Hub All Resources Change state

The other way to change the state is to select the offer settings and toggle the state, as shown in the following screenshot:

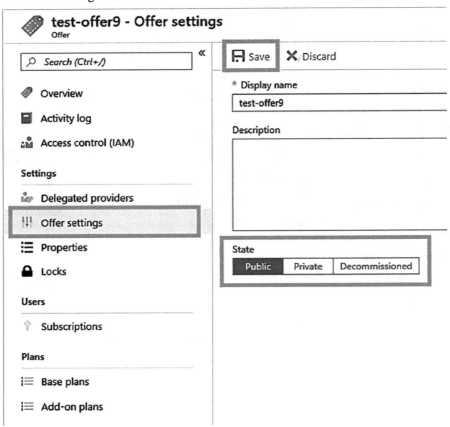

Figure 8.17 – Offer settings toggle state

Before we move on to looking at how users would subscribe to an offer, we will take a quick look at the add-on plans that can be used to modify a base plan when you want to offer additional services or extend compute, storage, or network quotas beyond the initial offer included with the base plan. Add-on plans are optional extensions that users can choose to enable with their subscription if they wish to extend beyond their base plan.

Add-on plans are created in the same manner as the base plan described in the previous section and the only difference is the point at which the plan is added to the offer. Add-on plans can be added to existing offers after the offer has been created to allow the offer to be extended.

Once an offer is created and made public, then it can be subscribed to, which we'll cover in the next section.

Subscribing to an offer

There are two ways for a user to subscribe to an offer:

- Cloud operators can create a subscription for a user from within the administrator portal. Subscriptions created by the cloud operator can include both public and private offers.

- Tenant users can subscribe to a public offer through the use of the user portal (self-service subscription creation).

We will work through each of these in turn, starting with the cloud operator subscription.

Cloud operator subscription

Cloud operators can use the administration portal to create a subscription to an offer on behalf of a user. Subscriptions can only be created for members of their own directory tenant. If multi-tenancy is enabled, you can also create subscriptions for users in the additional directory tenants.

The approach to creating subscriptions on behalf of tenants is a common approach when integrating Azure Stack Hub with external billing or service catalog systems. After a subscription has been created for a user, then when they next sign in to the user portal, they will see that they are subscribed to the offer.

The cloud operator will log in to the administrator portal and navigate to the user subscriptions.

They will then select **Add** and under **New user subscription**, provide the following information:

- **Display Name**: This is a friendly name to identify the subscription and will appear as the user subscription name.

- **User**: This is the specific user from the Active Directory tenant for this subscription. This username will appear as the owner. The format of the username is different depending on the identity manager:

  - **AAD**: `user1@contoso.onmicrosoft.com`

  - **ADFS**: `user1@azurestack.local`

- **Directory Tenant**: This is the directory that the user belongs to. If this is not a multi-tenancy system, then only the local directory is selectable.

Under **Offers**, choose the relevant offer for this subscription. As this subscription is being created on behalf of a user, then the accessibility state is set to **Private**.

Select **Create** and this will create the subscription. This subscription will then be visible under **User Subscription** and the user will see the subscription the next time they sign in to the user portal.

User subscription

As a user who is signed into the user portal, you can search for and subscribe to public offers for your directory tenant or organization.

Sign in to the user portal for Azure Stack Hub and select **Get a subscription** as shown in the following screenshot:

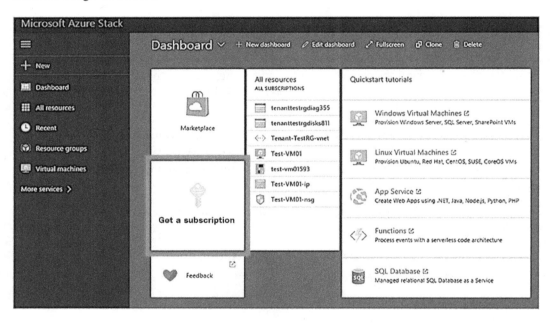

Figure 8.18 – Azure Stack Hub user portal Get a subscription

In the **Get a subscription** pane, enter a name for the subscription in the **Display name** field. Select **Offer** and then pick an offer from the **Choose an offer** pane as shown here:

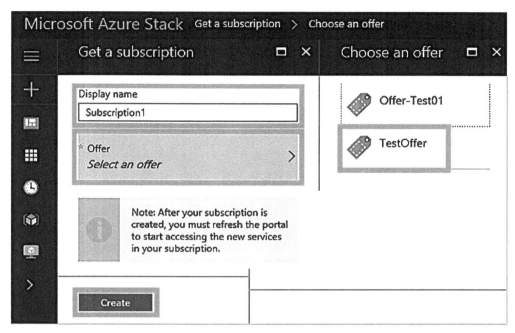

Figure 8.19 – Azure Stack Hub user portal Choose an offer

Selecting **Create** will then create the subscription. Once this has been created, refreshing the portal will then allow you to see the details of the services offered by the subscription. The subscription can be found in the **All Services** pane under the **General** category by selecting **Subscriptions**.

This covers the two different mechanisms for subscribing to offers and we'll now turn our attention to the delegation of offers.

Delegating offers

Delegation allows you to reach and manage more users than you can do by yourself. It allows you to put other people in charge of signing up users and creating subscriptions. This might be because you are a service provider who wants to let resellers sign up customers and manage them on your behalf. This could also equally apply to a central IT function delegating some responsibilities to individual departments.

Delegation can be shown using the structure in this diagram:

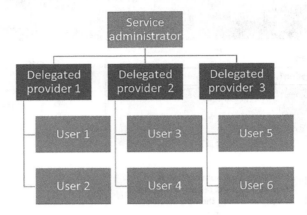

Figure 8.20 – Azure Stack Hub delegation

The service administrator is the cloud operator who manages the Azure Stack Hub infrastructure and is responsible for creating the offer template. The cloud operator then delegates to other users to provide others to their tenants. These are the delegated providers, and they will be users with either owner or contributor rights in the subscription. Users are then the tenant users who subscribe to the offers.

Delegation steps

There are two steps involved in setting up delegation:

- **Creation of a delegated provider subscription**: This is done by subscribing a user to an offer containing only the subscription service. Users subscribed to this offer can then extend the delegated offers to other users by signing them up.

- **Delegation of an offer to the delegated provider**: This allows the delegated provider to create subscriptions or to extend the offer to their users.

These steps are shown in the following diagram:

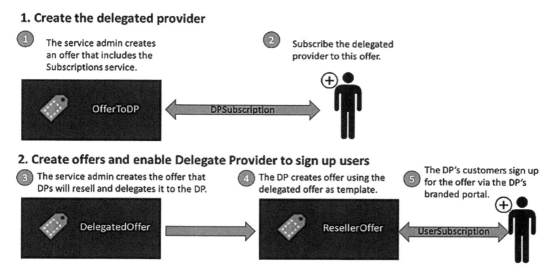

Figure 8.21 – Delegation steps

To be able to act as a delegated provider, a user must establish a relationship with the main provider by creating a subscription. This subscription identifies that the delegated provider has the right to present the delegated offers on behalf of the main provider. Once this relationship is established, the Azure Stack Hub operator can delegate an offer to the delegated provider. This then allows the delegated provider to take this offer and offer it to their customers. Delegated providers can only offer offers that have been delegated to them. They cannot make changes to these offers as only the Azure Stack Hub operator has the permissions to change these offers.

Creating an offer that allows a user to become a delegated provider begins with creating a plan and this plan only needs the subscription service. This is then followed by the creation of an offer that is based on this plan. Once the offer is created, then the user who will be the delegated provider can be subscribed to the offer.

This completes our coverage of the delegation process and we will now move on to the final section in this chapter where we look at managing plans and offers.

Managing plans and offers

In this final section of this chapter about plans and offers, we will look at the management of these plans and offers. It is possible to use PowerShell to configure and deliver services through the use of offers, plans, and subscriptions.

As a rule of thumb, you are not able to delete components that are not in use. As an example, you can only delete an offer when there are no subscriptions that are using that offer. The one exception to this is subscriptions. You can delete subscriptions and any resources that belong to this subscription will also be deleted. If you want to delete any of the other components, then you need to work backward and delete the components as you go. You would start with the offer, ensure that there are no subscriptions, and then delete the offer, followed by the plans and quotas.

Management of offers, plans, and subscriptions through the use of PowerShell relies on the Azure Stack Hub PowerShell module, which can be installed from a PowerShell console running as an administrator by using the `Import-Module AzureStack` cmdlet.

Quotas are needed when creating a plan. An existing quota can be used or you can create a new quota. For example, to create a new storage quota, you can use the `New-AzsStorageQuota` cmdlet in PowerShell:

```
$serviceQuotas = @()
$serviceQuotas += (New-AzsStorageQuota -Name "Example storage
quota with defaults").Id
$serviceQuotas += (New-AzsComputeQuota -Name "Example compute
quota with defaults").Id
$serviceQuotas += (New-AzsNetworkQuota -Name "Example network
quota with defaults").Id
```

To create a plan, use `New-AzsPlan` cmdlet as shown in this code example:

```
$testPlan = New-AzsPlan -Name "testplan" -ResourceGroupName
"testrg" -QuotaIds $serviceQuotas -Description "Test Plan"
```

An offer can then be created using the `New-AzsOffer` cmdlet as shown in this example:

```
New-AzsOffer -Name "testoffer" -ResourceGroupName "testrg"
-BasePlanIds @($testPlan.Id)
```

You can then add additional plans to the offer through the use of the
`Add-AzsPlanToOffer` cmdlet as shown in the next example:

```
Add-AzsPlanToOffer -PlanName "addonplan" -PlanLinkType
Addon -OfferName "testoffer" -ResourceGroupName "testrg"
-MaxAcquisitionCount 18
```

You can change the state of an offer by using the `Set-AzsOffer` cmdlet as shown here:

```
$offer = Get-AzsAdminManagedOffer -Name "testoffer"
-ResourceGroupName "testrg"
$offer.state = "Public"
$offer | Set-AzsOffer -Confirm:$false
```

Once the offer is created, then it is also possible to use PowerShell to subscribe to the offer
both as the cloud operator and as a user. To create the subscription on behalf of a user as
a cloud operator, then use the `New-AzsUserSubscription` cmdlet as shown here:

```
New-AzsUserSubscription -Owner "user@contoso.com" -DisplayName
"User subscription" -OfferId "/subscriptions/<Subscription ID>/
resourceGroups/testrg/providers/Microsoft.Subscriptions.Admin/
offers/testoffer"
```

To subscribe to the offer as a user, then the `New-AzsSubscription` cmdlet is used as
shown here:

```
$testoffer = Get-AzsOffer | Where-Object Name eq "testoffer"
New-AzsSubscription -DisplayName "User Subscription" -OfferId
$testoffer.Id -DisplayName "My Subscription"
```

There are equivalent `Remove` cmdlets for all the objects created via PowerShell, which are
shown here for reference:

```
Remove-AzsUserSubscription
Remove-AzsPlanFromOffer
Remove-AzsPlan
Remove-AzsOffer
Remove-AzsStorageQuota
Remove-AzsComputeQuota
Remove-AzsNetworkQuota
```

This completes our chapter on plans, offers, and subscriptions but before we move on to the next chapter about the Azure Stack Hub Marketplace, let's review what we have learned in this chapter.

Summary

In this chapter, we started by looking at the steps to plan the service offerings we want to offer to our users from Azure Stack Hub. We then looked at what it takes to create plans in Azure Stack Hub with their corresponding quotas. We then took these plans and used them to create offers. We looked at the different states, such as public and private, for these offers. We then went through the different options for subscribing users to these offers. We looked at the delegation of offers and the delegation provider responsibilities, and we then finished the chapter with a run-through of managing these objects using PowerShell.

This covers everything we need to know about plans, quotas, offers, and subscriptions. We will now move on to the next chapter on the Azure Stack Hub Marketplace.

9
Realizing Azure Marketplace

This chapter is focused on Microsoft Azure Marketplace and how it relates to Microsoft Azure Stack Hub. By the end of this chapter, you should be familiar with Azure Marketplace within Azure Stack Hub and be able to describe the different scenarios available for the different Azure Stack Hub connection states. You will learn how to create your own Marketplace items, as well as being able to utilize different solutions from vendors that have been published to Marketplace. You will learn how Marketplace in Azure and Azure Stack Hub are related. We will also touch on how to keep Marketplace in sync when you are running Azure Stack Hub in the disconnected scenario.

We will be walking through the following sections in this chapter:

- Overviewing Azure Stack Hub Marketplace
- Reviewing Azure Stack Hub Marketplace scenarios
- Creating and publishing items to Marketplace
- Understanding Azure Stack Hub Marketplace solution types

We will begin this chapter with an in-depth overview of Azure Marketplace.

Technical requirements

You can view this chapter's code in action here: `https://bit.ly/3jf9CGV`

Overviewing Azure Stack Hub Marketplace

Azure Stack Hub Marketplace is an online content store with a collection of services, apps, and resources that are customized for Azure Stack Hub. Resources include **virtual machines** (**VMs**), storage, and networking. It is possible to make use of Azure Stack Hub Marketplace to create new resources and deploy new applications or examine and pick items that you want to make use of. To be able to procure a Marketplace item, users must subscribe to an offer, which then grants them access to this item.

The Azure Stack Hub operator decides which items to add (publish) to Azure Stack Hub Marketplace. They can publish items such as databases, applications, app services, and more. By publishing items, these items then become visible to Azure Stack Hub users. Custom items that have been created by the Azure Stack Hub operator can be published alongside a growing list of Azure Marketplace items that are available. Any item that is published to Azure Stack Hub Marketplace is visible to users within 5 minutes.

Every item that is published to Azure Stack Hub Marketplace uses the Azure Gallery (`.azpkg`) format. Deployment or runtime resources (code, ZIP files with software, or VM images) should be added separately to Azure Stack Hub and not as part of the Marketplace item.

Azure Stack Hub converts any images to sparse files when they are downloaded from Azure or when a custom image is uploaded. This process increases the time taken to add an image but does save space and speeds up the deployment of the images. Existing images are not converted; this conversion is only applied to new images.

An item in Azure Stack Hub Marketplace is a service, app, or resource that Azure Stack Hub users can download and utilize. All of the items in Azure Stack Hub Marketplace are visible to all Azure Stack Hub users. This includes administrative items such as plans and offers. Administrative items do not require a subscription to be able to view them but will not be functional for most users.

Every Azure Stack Hub Marketplace item has the following characteristics:

- An **Azure Resource Manager** (**ARM**) template used to provision the resource
- Metadata, such as names, images, and other marketing material
- Formatting information to define how the item will be displayed in the portal

All items in the gallery, such as **JSON** files, are reachable without authentication after they have been made available in Azure Stack Hub Marketplace.

Items from Azure Marketplace are available for downloading to Azure Stack Hub, but only certain items are interchangeable between the two. This list is always evolving, and the current list can be found at the following URL: `https://docs.microsoft.com/en-us/azure-stack/operator/azure-stack-marketplace-azure-items?`.

Azure Stack Hub Marketplace is accessible from anywhere in the administrator portal and is opened by selecting **+ Create a resource**, as shown in the following screenshot:

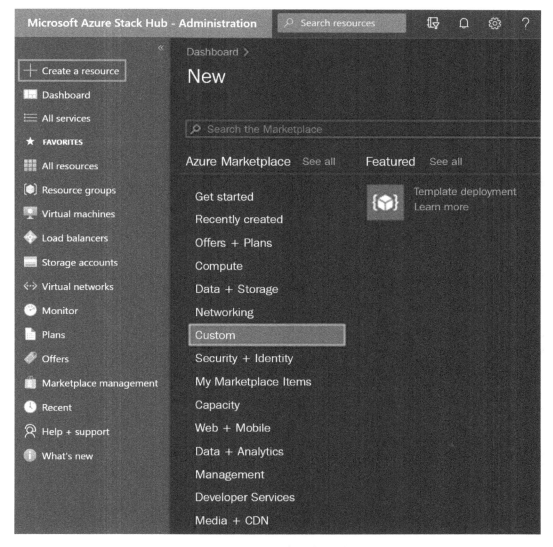

Figure 9.1 – Azure Stack Hub Marketplace

This completes our overview of Azure Stack Hub Marketplace and from here we will turn our attention to the different Azure Stack Hub Marketplace scenarios and what they mean for downloading items.

Reviewing Azure Stack Hub Marketplace scenarios

There are two distinct scenarios for downloading Azure Stack Hub Marketplace items:

- **Disconnected or partially connected**: Requires access to the internet through the use of the Marketplace syndication tool to download Marketplace items. The items are then transferred to the disconnected Azure Stack Hub installation. This uses PowerShell.

- **Connected**: The Azure Stack Hub environment is connected to the internet. The Azure Stack Hub administrator portal is used to find and download Marketplace items.

The catalog you see for the Marketplace items will be determined by the Azure subscription that was used for registering the Azure Stack Hub environment.

Let's walk through the process of downloading items from Marketplace in each scenario starting with the connected scenario.

Downloading Marketplace items to Azure Stack Hub in the connected scenario

Azure Stack Hub must already be registered with Azure and the solution must be running as a connected scenario rather than as a disconnected scenario.

Sign in to the Azure Stack Hub administrator portal at the following URL: `https://adminportal.local.azurestack.external`.

Before downloading new items from Marketplace, it is advisable to review the storage space available in Azure Stack Hub. This can be accomplished using **Region management | Resource providers | Storage**, as shown in the following screenshot:

Figure 9.2 – Azure Stack Hub available storage

To view the available items in Azure, select the Marketplace management service and then **Marketplace items | + Add from Azure,** as per the following screenshot:

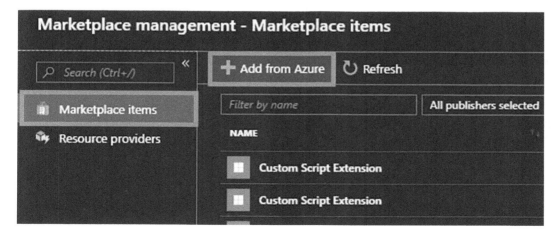

Figure 9.3 – Azure Stack Hub Marketplace management

Each line item that is listed details the currently available version. If there is more than one version of a particular Marketplace item available, then the **VERSION** column will display **Multiple**, as shown in the following screenshot:

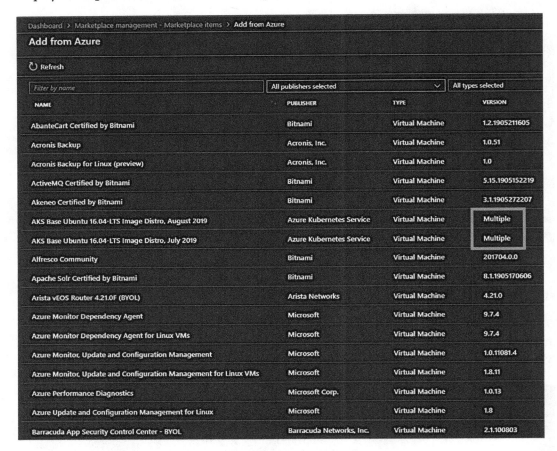

Figure 9.4 – Azure Stack Hub Marketplace items

If you select an item that shows **Multiple** in the **VERSION** column, then after selecting the item, you will be able to select the specific version from the **Version** dropdown, as shown in the following screenshot:

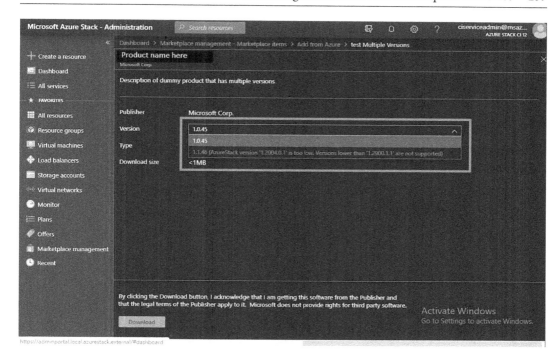

Figure 9.5 – Azure Stack Hub version selection

The item can be selected, and then click on **Download**. The time it takes to download the item will vary depending on the network connectivity. Once the item has been downloaded, you can deploy the new Marketplace item as either an Azure Stack Hub operator or an Azure Stack Hub user.

The item can be deployed by using **+ Create a resource** and locating the new Marketplace item. Once the new Marketplace item is selected, the deployment process is kicked off. This process will vary depending on the Marketplace item selected, as will the length of time taken. This process works well when the Azure Stack Hub environment is connected to the internet but for the disconnected scenario, the process is slightly different, which we will look at next.

Downloading Marketplace items to Azure Stack Hub in the disconnected scenario

If Azure Stack Hub has limited or no internet connectivity, you use PowerShell and the Marketplace syndication tool to download the Marketplace items to another machine that has internet connectivity. You will then transfer the downloaded items to the Azure Stack Hub environment. With a disconnected Azure Stack Hub, it is not possible to use the Azure Stack Hub portal to download the Marketplace items as we explained in the previous section. The Marketplace syndication tool can also be used when the Azure Stack Hub environment is connected and the administrator would rather use PowerShell.

There are two activities involved in downloading Marketplace items when in a disconnected environment:

- **Download from Marketplace**: On a machine that is connected to the internet, configure PowerShell, download the Marketplace syndication tool, and then download the items from the Marketplace.

- **Publish to Azure Stack Hub**: Once the files are downloaded, they are transferred to the Azure Stack Hub environment and then published to Azure Stack Hub Marketplace.

Prior to being able to utilize Marketplace from a disconnected Azure Stack Hub, the instance of Azure Stack Hub must already be registered. Then, the first step to being able to download Marketplace items to a disconnected Azure Stack Hub environment is to configure the PowerShell environment with the syndication admin. This can be installed from the PowerShell Gallery by issuing the following command in a PowerShell window running as an administrator:

```
Install-Module -Name Azs.Syndication.Admin -AllowPrerelease
-Passthru
```

Within the same PowerShell window, the following command will connect to Azure and the **Azure Active Directory** (**AAD**) tenant with the user who registered the Azure Stack Hub instance:

```
Connect-AzAccount -Environment AzureCloud -Tenant
'<mydirectory>.onmicrosoft.com'
```

This will prompt you to enter the Azure account credentials.

If you have more than one subscription, then the following command can be run in PowerShell to select the relevant subscription for Azure Stack Hub:

```
Get-AzSubscription -SubscriptionID 'Your Azure Subscription
GUID' | Select-AzSubscription
```

To be able to select Marketplace items to download, the following command is run in PowerShell:

```
$products = Select-AzsMarketplaceItem
```

This will display a list of the Azure Stack Hub registrations available based on the subscription logged into. Select the relevant Azure Stack Hub registration from this list as per the following screenshot:

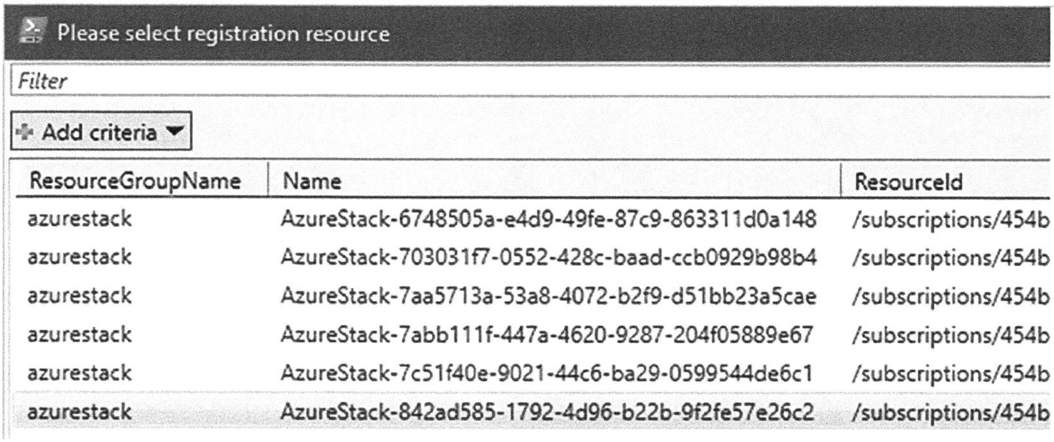
Please select registration resource
Filter
Add criteria
ResourceGroupName
Name
ResourceId
azurestack
AzureStack-6748505a-e4d9-49fe-87c9-863311d0a148
/subscriptions/454b
azurestack
AzureStack-703031f7-0552-428c-baad-ccb0929b98b4
/subscriptions/454b
azurestack
AzureStack-7aa5713a-53a8-4072-b2f9-d51bb23a5cae
/subscriptions/454b
azurestack
AzureStack-7abb111f-447a-4620-9287-204f05889e67
/subscriptions/454b
azurestack
AzureStack-7c51f40e-9021-44c6-ba29-0599544de6c1
/subscriptions/454b
azurestack
AzureStack-842ad585-1792-4d96-b22b-9f2fe57e26c2
/subscriptions/454b

Figure 9.6 – Azure Stack Hub registration selection

Once the relevant Azure Stack Hub registration has been selected, this will display a second list of the Marketplace items that are available for download, as shown in the following screenshot:

Name	Publisher	Type	Version	ResourceId
CloudLink SecureVM 6.6 BYOL	Dell EMC	virtualMachine	6.6.1	/subscriptions/454b15
CloudLink SecureVM 6.6 Solution BYOL	Dell EMC	virtualMachine	6.6.1	/subscriptions/454b15
CloudLink SecureVM 6.7 BYOL	Dell EMC	virtualMachine	6.7.1	/subscriptions/454b15
CloudLink SecureVM 6.7 Solution BYOL	Dell EMC	virtualMachine	1.0.0	/subscriptions/454b15
CloudLink SecureVM 6.8 BYOL	Dell EMC	virtualMachine	6.8.0	/subscriptions/454b15
CloudLink SecureVM 6.8 Solution BYOL	Dell EMC	virtualMachine	1.0.0	/subscriptions/454b15
CloudLink SecureVM 6.9 BYOL	Dell EMC	virtualMachine	6.9.0	/subscriptions/454b15
CloudLink SecureVM 6.9 Solution BYOL	Dell EMC	virtualMachine	1.0.0	/subscriptions/454b15
CloudLink SecureVM Agent	Dell EMC	virtualMachineExtension	6.8	/subscriptions/454b15
CloudLink SecureVM Agent	Dell EMC	virtualMachineExtension	6.8	/subscriptions/454b15
CloudLink SecureVM Agent for Linux	Dell EMC	virtualMachineExtension	6.0	/subscriptions/454b15
CloudLink SecureVM Agent for Windows	Dell EMC	virtualMachineExtension	6.5	/subscriptions/454b15
CMS Made Simple Certified by Bitnami	Bitnami	virtualMachine	2.2.1905210741	/subscriptions/454b15
Commvault Trial	Commvault	virtualMachine	11.13.1	/subscriptions/454b15
Composr Certified by Bitnami	Bitnami	virtualMachine	10.0.1905152221	/subscriptions/454b15
concrete5 Certified by Bitnami	Bitnami	virtualMachine	8.5.1905211806	/subscriptions/454b15
Coppermine Certified by Bitnami	Bitnami	virtualMachine	1.6.1905210742	/subscriptions/454b15
CoreOS Linux (Stable)	CoreOS	virtualMachine	1465.8.0	/subscriptions/454b15
CouchDB	Bitnami	virtualMachine	2.0.2	/subscriptions/454b15
CouchDB Certified by Bitnami	Bitnami	virtualMachine	2.3.1905152218	/subscriptions/454b15
Custom Script Extension	Microsoft Corp.	virtualMachineExtension	1.9.3	/subscriptions/454b15
Custom Script for Linux	Microsoft Corp.	virtualMachineExtension	1.5.2.2	/subscriptions/454b15
Custom Script for Linux 2.0	Microsoft Corp	virtualMachineExtension	2.0.6	/subscriptions/454b15
Data Box Edge/Data Box Gateway	Microsoft Corp.	resourceProvider	1.0.5	/subscriptions/454b15
Data Box Gateway Virtual Device	Microsoft	virtualMachine	1.0.2001	/subscriptions/454b15
Debian 8 "Jessie"	credativ	virtualMachine	8.0.20190806	/subscriptions/454b15
Debian 9 "Stretch"	credativ	virtualMachine	9.0.201805160	/subscriptions/454b15
Debian 9 "Stretch"	credativ	virtualMachine	9.0.201807160	/subscriptions/454b15
Dell EMC Data Domain Virtual Edition 3.1 - 6.1.0.X	Dell EMC	virtualMachine	6.1.0110	/subscriptions/454b15
Discourse Certified by Bitnami	Bitnami	virtualMachine	2.2.1905152216	/subscriptions/454b15
Django Certified by Bitnami	Bitnami	virtualMachine	2.2.1905160606	/subscriptions/454b15
DokuWiki	Bitnami	virtualMachine	201702192.0.0	/subscriptions/454b15
Dolibarr Certified by Bitnami	Bitnami	virtualMachine	9.0.1905160607	/subscriptions/454b15

Figure 9.7 – Azure Marketplace items for download

You can select the items you wish to download from the list. You can select multiple items to download by using the *Ctrl* key. It is possible to filter the list by making use of the + **Add criteria** button and entering criteria, as shown in the following screenshot:

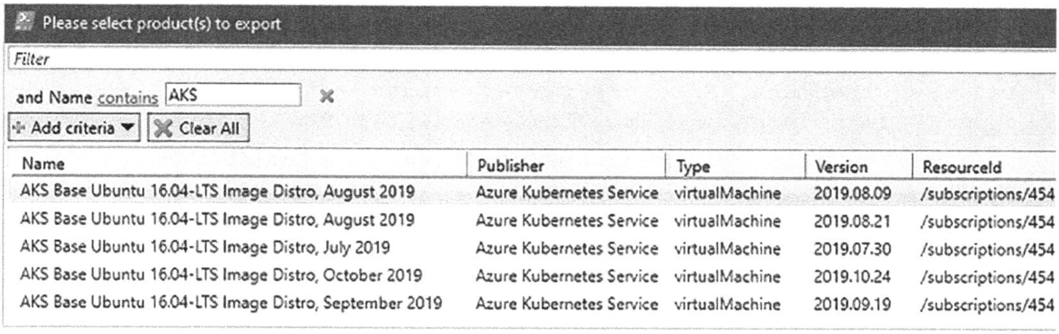

Figure 9.8 – Product filter criteria dialog

Once you have selected all of the items for download, click **OK**, which will then save the IDs of the Marketplace items selected into the $products variable. The following command can then be used in PowerShell to download the items to a folder of your choice:

```
$products | Export-AzsMarketplaceItem -RepositoryDir
"Destination folder path"
```

This download may take time depending on the number and type of items selected for download. If the download fails, then it can be rerun using the same command.

The Azs.Syndication.Admin module we installed earlier should also be exported so that it can be transferred to the Azure Stack Hub environment along with the Marketplace items. This module can be exported through the use of the following PowerShell cmdlet:

```
Save-Package -ProviderName NuGet -Source https://www.
powershellgallery.com/api/v2 -Name Azs.Syndication.Admin -Path
"Destination folder" -Force
```

The files that have been downloaded should now be copied across to another machine that has connectivity to the disconnected Azure Stack Hub instance. The Marketplace syndication tool should also be installed on this machine using the instructions from earlier in this section.

Use the following cmdlets from a PowerShell window running as administrator to connect to the Azure Stack Hub instance:

```
# Register an Azure Resource Manager environment that targets
your Azure Stack Hub instance. Get your Azure Resource Manager
endpoint value from your service provider.
   Add-AzEnvironment -Name "AzureStackAdmin" -ArmEndpoint
"https://adminmanagement.<region>.<externalFQDN>" '
      -AzureKeyVaultDnsSuffix adminvault.<region>.<externalFQDN>
'
      -AzureKeyVaultServiceEndpointResourceId https://
adminvault.<region>.<externalFQDN>

# Sign in to your environment.
Connect-AzAccount -EnvironmentName "AzureStackAdmin"
```

Once connected, use the following cmdlet to run the Marketplace syndication:

```
Import-AzsMarketplaceItem -RepositoryDir "source folder"
```

Once this script has finished, the Marketplace items should be visible in Azure Stack Hub Marketplace.

This completes the overview of the different processes required for accessing Marketplace items in the connected and disconnected scenarios. We will continue on our journey through Marketplace by looking at creating and publishing our own items to Marketplace.

Creating and publishing items to Marketplace

Items that are published to Azure Stack Hub Marketplace use the Azure Gallery Package format. The Azure Gallery Packager tool allows you to create a custom Azure Gallery package that can be uploaded to the Azure Stack Hub Marketplace for download by your users.

As an Azure Stack Hub operator, you can add your own custom VM images to Marketplace and make them available for download by your users. You can add the custom VM images to Azure Stack Hub Marketplace through the administrator portal or via PowerShell.

You have two options for enabling users to use an image:

- Offer the image via an ARM template.
- Offer the image through Azure Stack Hub Marketplace.

We have already covered ARM templates in *Chapter 8, Working with Offers, Plans, and Quotas*, of this book, so for this chapter, we will look at the ability to offer the image through Azure Stack Hub Marketplace.

Once an image has been added through the admin portal via **Compute | Image**, it can be referenced as the URI in the following process. A custom image can be generated from an existing VM or from Azure. Full instructions for doing this creation are available here: `https://docs.microsoft.com/en-us/azure-stack/user/vm-move-overview`.

To be able to add a custom VM image to Azure Stack Hub through the portal, start by logging in to the Azure Stack Hub admin portal as an operator. Select **Dashboard** and then **Compute** in the **Resource providers** panel, as shown in the following screenshot:

Figure 9.9 – Azure Stack Hub portal dashboard

From the **Compute** panel, select **VM images** and then select **+ Add**, as shown:

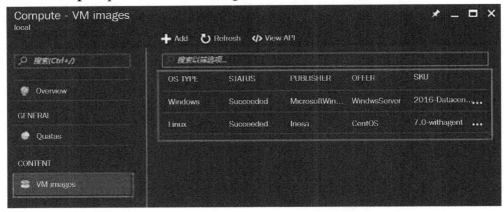

Figure 9.10 – Azure Stack Hub adding VM images

In the **Add a VM Image** panel, enter details for the **Publisher**, **Offer**, **OS Type**, **SKU**, **Version**, and **OS disk blob URI** fields, and then select **Create**. This will then create the new VM image, as shown here:

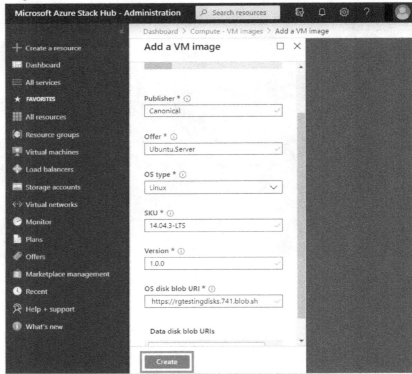

Figure 9.11 – Azure Stack Hub Add a VM image panel

Once the VM image has been created, it will be available to your users to work with through ARM but you can also make it available as a Marketplace item.

The Azure Gallery Packager tool is available for download from the following URL: `https://aka.ms/azsmarketplaceitem`.

This is downloaded as a ZIP file and once extracted contains sample packages and resource templates that can be modified to work for the custom VM image that was uploaded.

Use one of the sample templates or create a new template with the following structure:

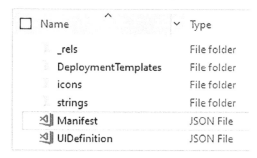

Figure 9.12 – Azure Stack Gallery package

An example `Manifest.json` file is shown as follows, and the numbered items will be updated based on the information used to create the custom VM image earlier:

```
{
    "$schema": "https://gallery.azure.com/schemas/2015-10-01/
Manifest.json#",
    "name": "Test", (1)
    "publisher": "<Publisher name>", (2)
    "version": "<Version number>", (3)
    "displayName": "ms-resource:displayName", (4)
    "publisherDisplayName": "ms-resource:publisherDisplayName",
(5)
    "publisherLegalName": "ms-resource:publisherDisplayName",
(6)
    "summary": "ms-resource:summary",
    "longSummary": "ms-resource:longSummary",
    "description": "ms-resource:description",
    "longDescription": "ms-resource:description",
    "links": [
```

```
      { "displayName": "ms-resource:documentationLink", "uri":
"http://go.microsoft.com/fwlink/?LinkId=532898" }
   ],
  "artifacts": [
    {
        "isDefault": true
    }
  ],
  "images": [{
    "context": "ibiza",
    "items": [{
        "id": "small",
        "path": "icons\\Small.png", (7)
        "type": "icon"
        },
        {
          "id": "medium",
          "path": "icons\\Medium.png",
          "type": "icon"
        },
        {
          "id": "large",
          "path": "icons\\Large.png",
          "type": "icon"
        },
        {
          "id": "wide",
          "path": "icons\\Wide.png",
          "type": "icon"
        }]
    }]
}
```

Each of the preceding fields that reference `ms-resource` will need to be updated in the equivalent `strings/resources.json` file, as shown in the following sample:

```
{
"displayName": "<OfferName.PublisherName.Version>",
"publisherDisplayName": "<Publisher name>",
"summary": "Create a simple VM",
"longSummary": "Create a simple VM and use it",
"description": "<p>This is just a sample of the type of
description you could create for your gallery item!</p><p>This
is a second paragraph.</p>",
"documentationLink": "Documentation"
}
```

Within the `DeploymentTemplates` folder, the structure appears as follows:

Figure 9.13 – Deployment templates structure

The values in the `createuidefinition.json` file should be replaced with the values used when uploading the custom VM image to Azure Stack Hub.

The ARM template should be saved to the deployment templates folder. Icons should be added to the `Icons` folder and text to the resources file in strings for the associated icons. Small, medium, large, and wide icons are required for building a Marketplace item correctly.

Once all of the files are modified, then to convert the gallery to the correct format, perform the following cmdlet in a PowerShell window:

```
.\AzureGalleryHubGallery.exe package -m c:\<path>\<gallery
package name>\manifest.json -o c:\Temp
```

The Marketplace item in the `azpkg` format should be copied to either a new Azure blob storage or local Azure Stack Hub storage, and then imported into your gallery using the following cmdlet:

```
Add-AzsGalleryItem -GalleryItemUri '
https://sample.blob.core.windows.net/<temporary blob
name>/<offerName.publisherName.version>.azpkg -Verbose
```

Once the gallery package has been imported, the custom VM image should be visible within Marketplace as well as the view for creating a new resource, as shown in the following screenshot:

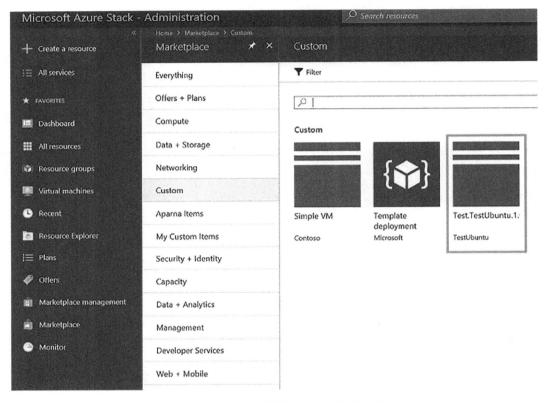

Figure 9.14 – Custom VM image in Marketplace

It is also possible to remove items from Marketplace via PowerShell by using the following cmdlet:

```
Remove-AzsGalleryItem -Name <Gallery package name> -Verbose
```

Now we know how to create and publish our own Marketplace items, so let's look at the different solution types available in Azure Stack Hub Marketplace.

Understanding Azure Stack Hub Marketplace solution types

Azure Stack Hub Marketplace effectively caters to four different solution types:

- Existing pre-built images and solutions
- Vendor VM images
- Vendor VM extensions
- Vendor pre-built solutions

Existing images are items that have been published by Microsoft or their partners to Marketplace to allow you to deploy resources such as Ubuntu and CentOS. The vendor VM images are as we described in the previous section, where you may want to upload your own custom image. This may be because you have a standard configured template you use for your Windows deployments, for example.

Azure VM extensions can apply post-deployment configuration and automation to VM images. This can include things such as post-install software deployment or Docker configuration or perhaps anti-virus protection. They can be run in PowerShell, through the Azure portal, or initiated as part of an ARM template. They can be bundled with a new VM deployment or run against any existing system. From an Azure Stack Hub standpoint, the extensions are treated no differently than the VM images.

A solution is a fit-and-finish solution that provides a platform or service on OS images and third-party software. It can be multi-image or single-image based. It can also be a mix of Azure Stack Hub services such as a VM scale set, a load balancer, a storage account, and a key vault all interacting together as a solution. It is designed by an **independent software vendor** (**ISV**) and syndicated via Azure Marketplace. There are also solutions that customers can create and publish themselves.

A solution includes everything that is needed to run, including ARM templates, VM extensions, artifacts, and gallery packages.

Before we move on, there is one other offering within the Azure Stack Hub Marketplace that needs consideration, and that is Kubernetes.

Offering a Kubernetes marketplace item

To be able to offer Kubernetes as a service from your Azure Stack Hub instance to allow your users to deploy a Kubernetes cluster, first you will need to set up the **Azure Kubernetes Service (AKS)** engine. The AKS engine is an ARM template-driven way to provision a self-managed Kubernetes cluster on Azure Stack Hub. As this is not a managed Kubernetes solution, there is limited support for ongoing operational capabilities such as scaling, in-place upgrade, and extensions that you would find in a fully managed Kubernetes solution.

Users who wish to use the service will need a plan, offer, and subscription to Azure Stack Hub with enough space as users will often deploy clusters of up to six VMs, which is made up of three masters and three worker nodes. They will need sufficient quotas to be able to accomplish this.

The Kubernetes cluster is reliant on a service principal and role-based permissions in Azure Stack Hub. We covered the creation of service principals in both **AAD** and **Active Directory Federation Services (ADFS)** in *Chapter 4, Exploring Azure Stack Hub Identity,* when we talked about identity.

You will need to download the AKS base image from Marketplace using the instructions within this chapter depending on the connected or disconnected state of the Azure Stack Hub instance.

You will also need to download a custom script for Linux from Azure and add it to the Azure Stack Hub Marketplace.

The AKS engine is a command-line tool to bootstrap Kubernetes clusters on Azure and Azure Stack Hub. Through the use of ARM, the AKS engine helps to create and maintain clusters running on VMs, virtual networks, and other infrastructure-as-a-service resources in Azure Stack Hub.

You can also offer Kubernetes as a Marketplace item as a standalone Kubernetes cluster, which is suitable for proofs of concept when the full AKS experience is not required.

This, again as with the AKS engine, is reliant on a service principal being created and role-based permissions in Azure Stack Hub.

This time, an Ubuntu image should be downloaded from Azure Marketplace and published to Azure Stack Hub alongside the custom script for Linux.

The Kubernetes cluster can then also be added from Azure to the Azure Stack Hub Marketplace using the following steps:

1. Navigate to the administrator portal at the following URL: `https://adminportal.local.azurestack.external`.

2. Select **All Services** and then **Marketplace Management** under the **Administration** category.

3. Select **+ Add from Azure**, enter `Kubernetes`, and then select **Kubernetes Cluster** followed by **Download**:

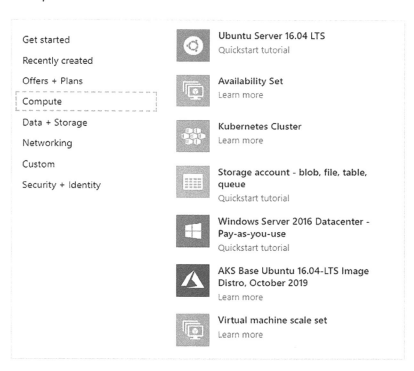

Figure 9.15 – Azure Stack Hub Kubernetes

This may take a few minutes to complete before the Marketplace item appears in Azure Stack Hub.

This completes our coverage of Azure Stack Hub Marketplace and before we move on to the next chapter on virtual networking, let's remind ourselves of what we have learned.

Summary

In this chapter, we have covered the Azure Stack Hub Marketplace and how it can be used to offer solutions and services to our users. We have covered how to populate Azure Stack Hub Marketplace in both the connected and disconnected scenarios, including the use of the Marketplace syndication tool. We have walked through creating our own custom images for publishing to Azure Stack Hub Marketplace and then also looked at the options for publishing Kubernetes to Azure Stack Hub Marketplace.

We will now move on to take a closer look at virtual networking in the next chapter.

10
Interpreting Virtual Networking

In this chapter, we will start to look at the concept of **Infrastructure-as-a-Service (IaaS)** within Microsoft Azure Stack Hub, beginning with virtual networking. We will cover each of the components included within virtual networking in detail to provide you with the knowledge you need to administer virtual networking within Microsoft Azure Stack Hub. This will include details on the different network services, such as DNS services. We will walk through the different connectivity options for the network. We will also look through the different IP address ranges needed for networking in Azure Stack Hub. By the end of this chapter, you should have a good understanding of virtual network components such as peering.

In this chapter, we will be covering the following topics:

- Understanding the software-defined network architecture
- Reviewing network services
- Understanding DNS services
- Introducing network connectivity

We will begin this chapter by understanding the software-defined network architecture in Azure Stack Hub.

Understanding the software-defined network architecture

Azure Stack Hub network resource creation is consistent with Azure, as it brings the power of the new Microsoft software-defined networking stack to Azure Stack Hub. Azure Stack Hub's **software-defined networking** (**SDN**) embraces industry standards such as **Virtual Extensible LAN** (**VXLAN**) and **Open vSwitch Database** (**OBVSB**), and incorporates technologies directly from Azure such as the **software load balancer** (**SLB**) and **Virtual Filtering Platform** (**VFP**) vSwitch extension, which is proven to be reliable, robust, and scalable.

Azure Stack Hub SDN leverages investments in Windows 2019 such as Packet Direct, **switch embedded teaming** (**SET**), converged NIC, offloads, and many others to provide software-defined networking capabilities at cloud scale.

Each component in Azure Stack Hub has its own set of **virtual IP** (**VIP**) addresses. The following diagram shows this, with each arrow representing a set of VIPs in Azure Stack Hub:

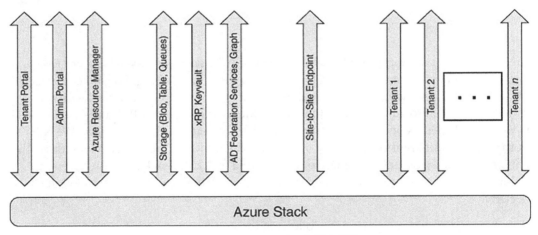

Figure 10.1 – Azure Stack Hub VIP

Azure Stack Hub will deploy a switch embedded teaming on top of the two physical network cards and will create five NICs:

- Two for the storage network.

- One for managing the infrastructure network.

- Two for the provider address's **Hyper-V network virtualization** (**HNV**) compartment, where the packets will be encapsulated.

The network controller, ADs, SLBs, MUXes, SDN GWs, and other infrastructure VMs will be connected to the infrastructure network.

The tenant's virtual networks will be defined on top of the HNV compartment, as shown in the following diagram:

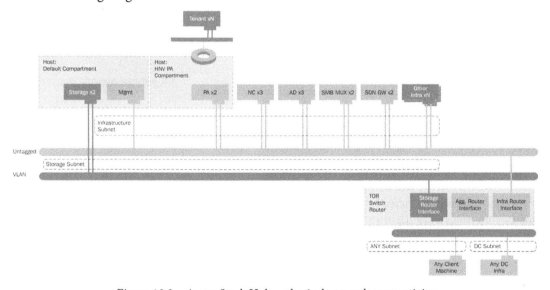

Figure 10.2 – Azure Stack Hub – physical network connectivity

For internal load balancing purposes, Azure Stack Hub defines a private VIP pool. This pool will be used by the software load balancer manager and will provide load balancing between subnets in the same virtual network. This is shown in the following diagram:

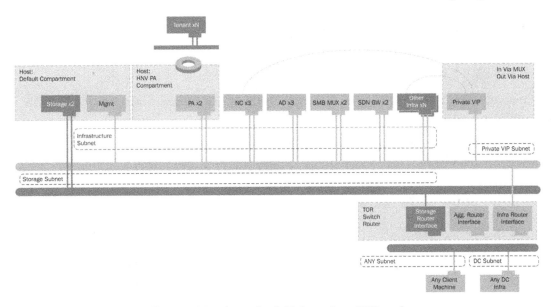

Figure 10.3 – Azure Stack Hub – private VIP pool

For public load balancing purposes, Azure Stack Hub defines a public VIP pool. This pool contains the accessible external or public IP addresses that will be assigned to a small set of Azure Stack Hub services, with the remainder to be used by the tenant VMs. This is shown in the following diagram:

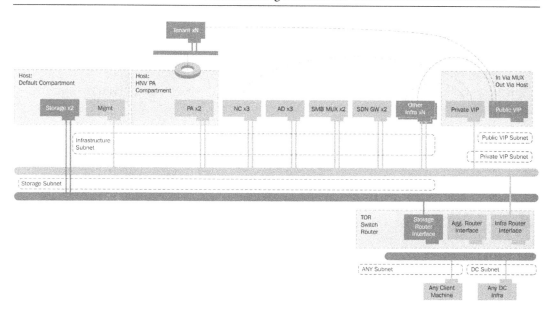

Figure 10.4 – Azure Stack Hub – full network connectivity

Azure Stack Hub builds upon the Windows Server 2019 SDN implementation. By implementing SDN, you can configure both the physical and virtual network devices. The core elements of Azure Stack Hub's SDN infrastructure include Hyper-V Network Virtualization and RAS gateway. Although the existing physical switches, routers, and other hardware devices can still be used independently, deeper integration can be achieved between the virtual network and the physical network in Azure Stack Hub, since these devices are designed for compatibility with software-defined networking.

Implementing SDN in Azure Stack Hub is possible since the three network planes – the management, control, and data planes – are independent of the network devices themselves, but they are abstracted for use by other entities such as Azure Stack Hub. SDN allows Azure Stack Hub to dynamically manage its data center network to provide an automated, centralized mechanism to meet the requirements of the applications and workloads. SDN in Azure Stack Hub provides the following capabilities:

- The ability to abstract applications and workloads from the underlying physical network, by virtualizing the network. Just as with server virtualization using Hyper-V, abstractions are consistent and work with the applications and workloads in a non-disruptive manner. In fact, software-defined networking provides virtual abstractions for Azure Stack Hub's physical network elements, such as IP addresses, switches, and load balancers.

- The ability to centrally define and control policies that govern both physical and virtual networks, including traffic flow between these two network types.

- The ability to implement network policies consistently at scale, even as you deploy new workloads or move workloads across the virtual or physical networks in Azure Stack Hub.

The SDN architecture in Azure Stack Hub is designed to provide hyper-scale networking capacities, similar to Azure. In the management plane, ARM handles authentication, authorization, and orchestration and, in turn, communicates directly with the **network resource provider** (**NRP**). NRP, like other foundational resource providers in Azure Stack Hub, is a web service that provides a foundation for all Azure Stack Hub-based IaaS and PaaS services. ARM relies on different resource providers to provide access to Azure Stack Hub Services. This Azure Stack Hub SDN architecture is shown in the following diagram:

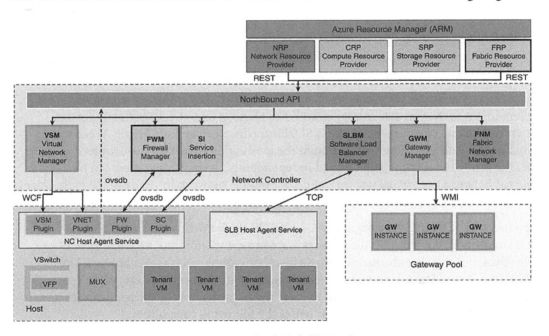

Figure 10.5 – Azure Stack Hub SDN architecture

Now that we have an understanding of the software-defined network architecture within Azure Stack Hub, we can move on and look at the network services.

Reviewing network services

In this chapter, we are going to look at the network services that are available within Azure Stack Hub. Azure Stack Hub includes multiple IaaS network resources, including the following:

- Virtual networks
- Subnets
- **Network security groups** (**NSGs**)
- Network security rules
- **Network interface** (**NICs**)
- **User-defined routes** (**UDRs**)
- Load balancers
- Public IP addresses
- DNS zones
- Virtual network gateways
- Local network gateways
- Connections

We will cover some of these network services in turn, starting with virtual networks and subnets.

Virtual networks and subnets

Virtual networks are completely segregated from one another, allowing you to create divided networks for development, testing, and production that all use the same **classless inter-domain routing** (**CIDR**) address blocks. You can then create subnets with your private or public IP address spaces. Internet access is provided through **network address translation** (**NAT**) by default, with no gateway required, as shown in the following diagram:

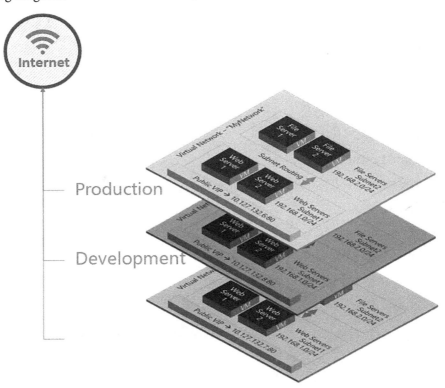

Figure 10.6 – Azure Stack Hub virtual network

Moving on from virtual networks and subnets, we will now look at network security groups.

Network security groups

Network security groups can be used to segment the network and can help protect internal traffic. They can be used to enable DMZ subnets and are associated with subnets and NICS. ACLs can be updated independently of VMs. An example is shown in the following diagram:

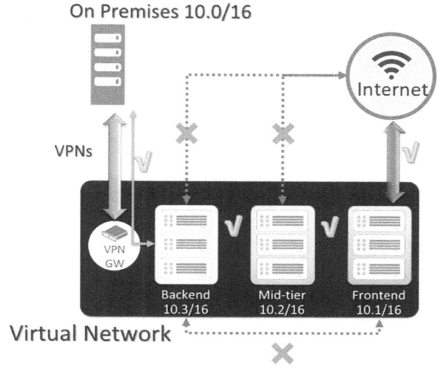

Figure 10.7 – Azure Stack Hub network security groups

In this diagram, we have defined multiple NSGs, such as the NSG for the frontend 10.1/16 space to allow access from the internet. This includes the following components:

- Priority 100
- Source internet
- Source ports *
- Destination frontend
- Destination port 80
- Protocol TCP
- Access allowed

The NSG for the backend and mid-tier would be similar, but the access would be set to Deny rather than Allow.

The following screenshot shows an example of adding an inbound security rule to a security group:

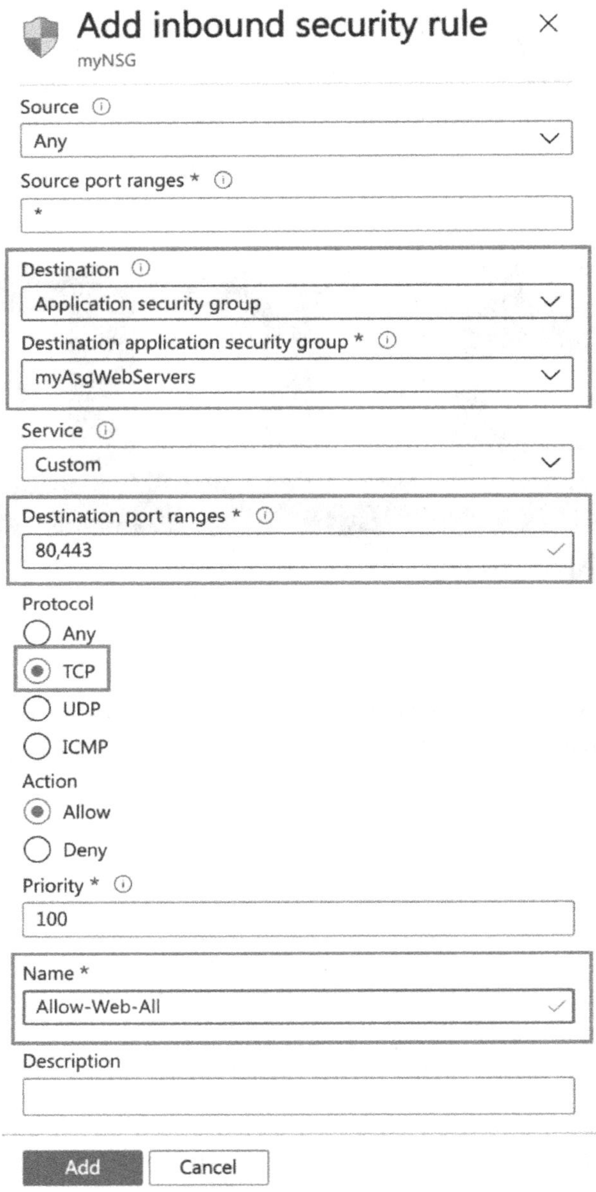

Figure 10.8 – Azure Stack Hub inbound security rule

Now that we've covered network security groups, let's review user-defined routes.

User-defined routes

User-defined routes allow you to control traffic flow with custom routes. This allows you to attach route tables to subnets. You can specify the next hop for any address prefix and set a default route to tunnel all traffic to on-premises or virtual appliances, as shown in the following diagram:

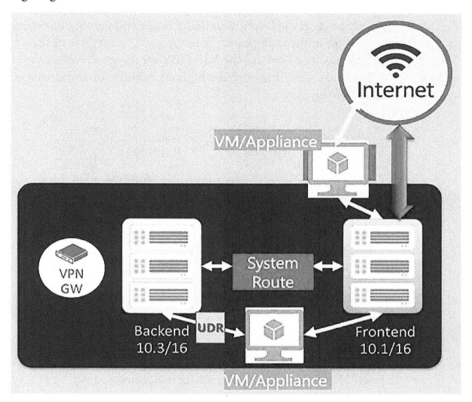

Figure 10.9 – Azure Stack Hub user-defined routes

Now that we've looked at user-defined routes, let's look at the software load balancer.

Software load balancer

The software load balancer helps balance loads among one or more VM instances. It is a Layer 4 load balancer, which allows port redirection. You can control traffic to endpoints with inbound **NAT** rules. It works by mapping **virtual IP addresses (VIPs)** to DIPs that are part of a cloud service set of resources inside the data center. VIPs are single IP addresses that provide public access to a pool of load-balanced VMs. For example, VIPs are IP addresses that are exposed on the internet so that tenants and tenant customers can connect to tenant resources in the data center. DIPs are the IP addresses of the member VMs of a load-balanced pool behind the VIP. DIPs are assigned within the cloud infrastructure to the tenant resources. This provides high availability for workloads with failover, as shown in the following diagram:

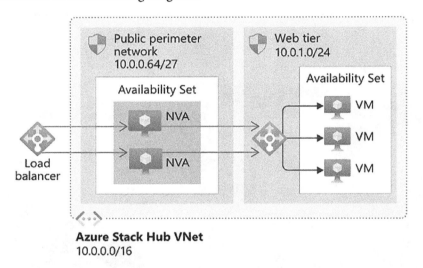

Figure 10.10 – Azure Stack Hub software load balancer

Public IP addresses can be assigned to load balancers and VMs, but only IPv4 is supported.

VIPs are located in the SLB **Multiplexer (MUX)**. The MUX consists of one or more VMs. The network controller provides each MUX with each VIP, and each MUX, in turn, uses a **Border Gateway Protocol (BGP)** to advertise each VIP to routers on the physical network as a /32 route. BGP allows the physical network routes to do the following:

- Learn that a VIP is available on each MUX, even if the MUXes are on different subnets in a Layer 3 network.

- Spread the load for each VIP across all available MUXes using **equal-cost multi-path (ECMP)** routing.

- Automatically detect a MUX failure or removal and stop sending traffic to the failed MUX.

- Spread the load from the failed or removed MUX across the healthy MUXes.

When public traffic arrives from the internet, the SLB MUX examines the traffic, which contains the VIP as a destination and maps and rewrites the traffic so that it will arrive at an individual DIP. For inbound network traffic, this transaction is performed in a two-step process that is split between the MUX VMs and the Hyper-V host where the destination DIP is located:

1. **Load balance**: The MUX uses the VIP to select a DIP, encapsulates the packet, and forwards the traffic to the Hyper-V host where the DIP is located.

2. **Network Address Translation (NAT)**: The Hyper-V host removes encapsulation from the packet, translates the VIP into a DIP, remaps the ports, and forwards the packet to the DIP VM.

The MUX knows how to map VIPs to the correct DIPs because of load balancing policies that you define by using a network controller. These rules include the protocol, frontend port, backend port, and the distribution algorithm.

When tenant VMs respond and send outbound network traffic back to the internet or remote tenant locations, because NAT is performed by the Hyper-V host, traffic bypasses the MUX and goes directly to the edge router from the Hyper-V host. After the initial network traffic has been established, the inbound network traffic also bypasses the SLB MUX completely.

From here, we will move on to the next section, where we will cover DNS services.

Understanding DNS services

Azure Stack Hub supports DNS hostname resolution and allows you to create and manage DNS zones and records using the API.

Tenants can create a DNS zone in Azure Stack Hub and create DNS records in that zone. Zones and records are resources that belong to the tenant subscription that created them. These resources can only be modified and deleted by that subscription. Unlike Azure, however, any tenant can resolve queries for these records. It should be noted that, unlike Azure DNS, Azure Stack Hub DNS is not multi-tenant.

The `<region>` DNS zone provides name resolution for the infrastructure virtual machines such as the network controller, the domain controllers, the load balancers, and the gateways. It is also the internal domain name used by the Azure Stack Hub infrastructure.

The `internal` DNS zone provides the following capabilities:

- Shared DNS name resolution services for a tenant workload's **internal domain name service (iDNS)**.

- An authoritative DNS service for name resolution and DNS registration within the tenant virtual network.

- A recursive DNS service for the resolution of internet names from tenant VMs. Tenants no longer need to specify custom DNS entries to resolve internet names.

- Each virtual network has its own GUID DNS subzone folder. The GUID represents the VNET resource ID that was created in the network controller by Azure Stack Hub.

The `<region>.cloudapp.<externalfqdn>` DNS zone is used to expose tenant applications or services to the internet or other external networks using public IP addresses defined in Azure Stack Hub.

The DNS service provided by Azure Stack Hub is similar to the DNS service in Azure, in that it uses APIs that are consistent with the Azure DNS. Hosting the domains in Azure Stack Hub DNS means you can manage the DNS records using the same credentials, APIs, and tools.

The Azure Stack Hub DNS service is more compact compared to the equivalent service in Azure. Azure Stack Hub DNS's scope, scale, and performance are affected by the size and location of the Azure Stack Hub deployment. Performance, availability, high availability, and global distribution will vary from Azure Stack Hub deployment to Azure Stack Hub deployment based on its size and location.

DNS forwarders in Azure Stack Hub DNS servers allow infrastructure servers and tenant virtual machines to resolve names from zones within your data center.

Azure Stack Hub DNS supports the use of ARM tags within the DNS zone's resources and supports up to 100 zones per subscription. It also supports up to 5,000 record sets within each zone and a maximum of 20 records per record set.

Azure Stack Hub DNS specifies records through the use of relative names. The fully qualified domain name includes the zone name, while the relative record does not. Record sets are used when a relative name and type is allocated to more than one IP address, such as a website. DNS record sets are effectively a collection of DNS records in a zone with the same name and type.

Now, let's look at the network connectivity that's offered in Azure Stack Hub, including virtual private networks, gateways, and virtual network peering.

Introducing network connectivity

We will begin this section on network connectivity by taking a look at the virtual private network gateway.

Virtual private network gateway

A **virtual private network** gateway, or **VPN** gateway, is a gateway that sends encrypted traffic across a public internet connection. This can be used to securely send traffic between virtual networks in Azure Stack Hub and corresponding virtual networks in Azure. Azure Stack Hub currently only supports one type of virtual network gateway, which is this VPN gateway.

A VPN gateway connection depends on resources with specific configuration settings. A lot of these resources can be configured in any order, while others must be configured in a certain order.

When connecting from Azure Stack Hub to external resources, you have two options available to you, as follows:

- **Site-to-Site (S2S)** VPN
- Outbound NAT

An S2S VPN connection requires a VPN device or **routing and remote access service (RRAS)**. The connection is encrypted and secure and runs across IPsec using an **internet key exchange (IKEv1 or IKEv2)**. This standard S2S VPN is shown in the following diagram:

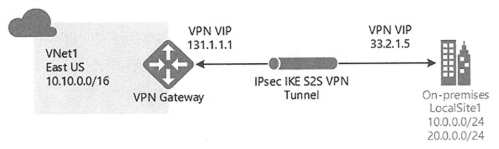

Figure 10.11 – Azure Stack Hub S2S VPN

There is another variation of the S2S VPN that allows more than one VPN connection from the Azure virtual network gateway. This is a multi-site VPN connection and is normally used for connecting to more than one on-premises site, as shown in the following diagram:

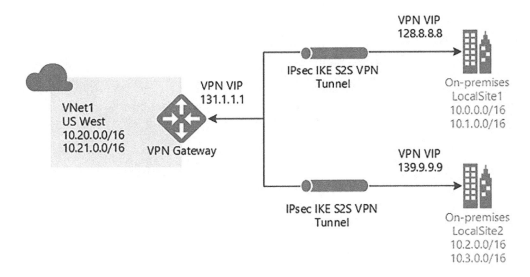

Figure 10.12 – Azure Stack Hub multi-site VPN

When running a multi-site VPN, since each virtual network can only have one VPN gateway, they will share the bandwidth between them.

Azure Stack Hub does not currently support policy-based gateways.

Azure Stack Hub supports the following VPN gateway SKUs, which are selected when the VPN gateway is created:

- Basic

- Standard

- High performance

The higher the level of SKU selected, the more CPUs and network bandwidth is allocated for the gateway. The more resources that are allocated, the higher the network throughput to the virtual network that's supported. For high availability scenarios, only the high-performance SKU can be used; Azure Stack Hub only supports active/passive configuration.

It should be noted that if you're using S2S VPN between two Azure Stack Hub deployments, then only one S2S VPN can be created.

It is possible to configure an IPsec/IKE policy for a S2S VPN gateway, but only against the standard or high-performance SKUs, since this is not supported for basic SKUs. This allows you to select different settings, such as encryption and integrity, for the cryptographic algorithms, such as the following:

- **IKEv2 encryption**:

  AES256, AES192, AES128, DES3, DES

- **IKEv2 integrity**:

  SHA384, SHA256, SHA1, MD5

- **DH group**:

  ECP384, DHGroup14, DHGroup2, DHGroup1, ECP256, DHGroup24

- **IPsec encryption**:

  GCMAES256, GCMAES192, GCMAES128, AES256, AES192, AES128, DES3, DES, None

- **IPsec integrity**:

  GCMAES256, GCMAES192, GCMAES128, SHA256

- **PFS group**:

  PFS24, ECP384, ECP256, PFS2048, PFS2, PFS1, PFSMM, None

Before you create a VPN gateway, you must create a gateway subnet. The gateway subnet contains the IP addresses that the virtual network gateway VMs and services use. When you create your virtual network gateway, gateway VMs are deployed to the gateway subnet and configured with the required VPN gateway settings.

When you create the gateway subnet, you specify the number of IP addresses that the subnet contains. The IP addresses in the gateway subnet are allocated to the gateway VMs and services. Some configurations require more IP addresses than others.

When creating a VPN gateway in Azure, the local network gateway often represents your on-premises network. In Azure Stack Hub, it represents any remote VPN device that sits outside Azure Stack Hub.

The last piece to look at with regards to the networking services within Azure Stack Hub is virtual network peering, which we will look at now.

Virtual network peering

Virtual network peering facilitates the ability to be able to seamlessly connect virtual networks in the Azure Stack Hub solution. From a connectivity viewpoint, the virtual networks will appear as a single network, and the traffic between the virtual machines will use the underlying **software-defined networking (SDN)** infrastructure. This means that this traffic behaves the same as the traffic between the virtual machines on the same virtual network. This keeps the traffic between these peered virtual networks private, which means no public internet, gateways, or encryption are required for this traffic to travel across. Network security groups can also be applied to these peered virtual networks.

User-defined routing can be used to enable service chaining, which allows you to direct traffic from a virtual network to a virtual appliance or gateway. This can be enabled by configuring the next hop to a VM in the peered network.

Service chaining allows you to deploy hub-and-spoke networks, where the network virtual appliance or VPN gateway is hosted on the hub virtual network. This then enables the spoke virtual networks to peer with the hub virtual network.

Every virtual network, including the peered virtual network, can have a gateway. This gateway can then be used to connect to on-premises networks. With a peered network, the gateway can also be configured as a transit point to on-premises networks. In this instance, the virtual network that is using a remote gateway is unable to have its own gateway as a virtual network – it is only able to have one gateway. This is shown in the following diagram:

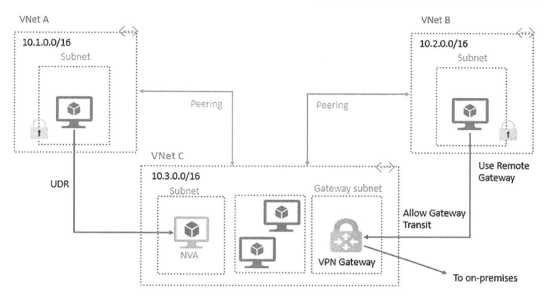

Figure 10.13 – Azure Stack Hub virtual gateways

Several checkbox settings can be configured when setting up virtual network peering, including the following:

- **Allow virtual network access**: The enables communication between virtual networks to allow connected resources to communicate to each other, as if they were on the same virtual network.

- **Allow forwarded traffic**: This enables traffic that's been forwarded by a network virtual appliance in another virtual network to flow to this peered virtual network.

- **Allow gateway transit**: This enables traffic to flow from the peered network through the gateway.

- **Use remote gateways**: This enables traffic from the virtual network to go through a gateway attached to the peered network.

When creating virtual network peering between different subscriptions and AAD tenants, the accounts being used will have the contributor role assigned to them. As there is no user interface for peering between AAD tenants, either PowerShell or the Azure CLI must be used.

To be able to allow the users of Azure Stack Hub to be able to create network solutions, including virtual private networks and virtual network peering, then, as an Azure Stack Hub operator, you can add a FortiGate **Next-Generation Firewall** (**NGFW**), or another **network virtual appliance** (**NVA**), from the marketplace.

With that, we have looked at the virtual networking components of Azure Stack Hub. Before we move on to the next chapter, where we will cover storage and compute features, let's review what we have learned in this chapter.

Summary

In this chapter, we started by understanding the virtual network architecture with Azure Stack Hub. This included physical network connectivity and the logical networks. From there, we ran through the different network services, such as software load balancers, network security groups, network peering, and DNS services. We then took a deeper dive into the DNS services and the different types of DNS servers available. From there, we moved on to network connectivity, including virtual private network gateways. Finally, we ran through virtual network peering.

In the next chapter, we will cover storage and compute resources.

11
Grasping Storage and Compute Fundamentals

This chapter covers the final two components of the **Infrastructure-as-a-Service** (**IaaS**) piece within Microsoft Azure Stack Hub. We will begin by looking at the available storage components and provide an overview of storage, including storage types, storage usage, and storage management. Finally, we will cover the final component of IaaS, which is the compute component within Microsoft Azure Stack Hub. This will introduce us to virtual machines, managed disks, and controllers.

We will be covering the following topics in this chapter:

- Overviewing Azure Stack Hub Storage
- Reviewing Azure Stack Hub Storage management and usage
- Overviewing Azure Stack Hub compute
- Understanding Azure Stack Hub compute components
- Reviewing Azure Stack Hub compute scenarios

We will begin by providing an in-depth overview of Azure Stack Hub Storage.

Technical requirements

You can view this chapter's code in action here: `https://bit.ly/3ya4E2F`

Overviewing Azure Stack Hub Storage

Azure Stack Hub Storage services are underpinned by the storage account that sits on top of the Azure Stack Hub infrastructure. The storage services that sit on top of this storage account can be split between IaaS and **Platform-as-a-Service (PaaS)**. These services include the following components:

- Disks
- Objects
- Tables
- Queues

These services help support other services such as virtual machines, custom apps, and microservices, as shown in the following diagram:

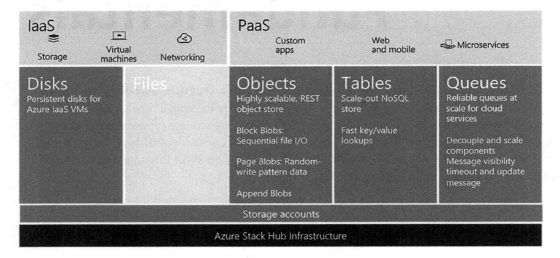

Figure 11.1 – Azure Stack Hub Storage services

These Azure Stack Hub Storage services are uniform with the storage services that are supplied by Azure storage in the public cloud.

The Azure Stack Hub Storage account is a secure account that allows us to access the storage services within Azure Stack Hub. The storage account is responsible for the unique namespace of the storage resources.

From an operator's standpoint, the storage services can be viewed as per the following diagram:

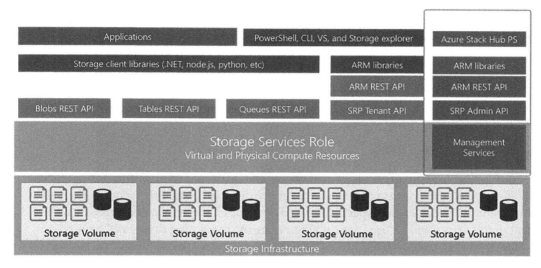

Figure 11.2 – Azure Stack Hub Storage services – operator view

These same services, when viewed from the tenant viewpoint, look as follows:

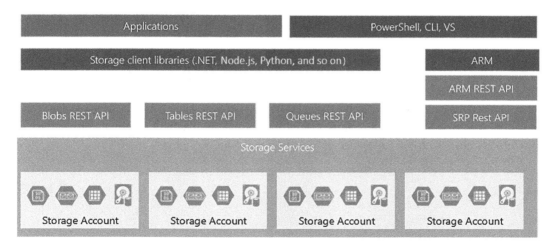

Figure 11.3 – Azure Stack Hub Storage services – tenant view

Before we walk through the storage services, we need to look at the infrastructure that underpins all of these services, starting with the drives.

Drives

The Azure Stack Hub integrated systems that are offered by the **Original Equipment Manufacturer (OEM)** vendors come in many different forms when it comes to the drive configuration. An integrated system can be configured with up to two different drive types out of the three supported drive types in Azure Stack Hub. The supported drive types are as follows:

- **Non-volatile memory express (NVMe)**
- **Solid state drive (SSD)/Serial Advanced Technology Attachment (SATA)/Serial Attached SCSI (SAS)**
- **Hard disk drive (HDD)**

The configuration of the drive's type affects how the drives are used within Azure Stack Hub. If only a single drive type is chosen from the three supported drive types, then all of the drives will be used for capacity. If two drive types have been configured, then the fastest of the drives will be used for caching, while the remaining drives will be used for capacity.

When selecting a solution from an OEM vendor, these can be grouped based on the drive configuration; that is, either *all-flash* or *hybrid*. Typically, an all-flash configuration is used to maximize storage performance and would not include any HDDs. A hybrid deployment would be used to balance performance and capacity and may include HDD drives.

The cache behavior is automatically configured based on the types of drives selected in the configuration. The main difference in caching is the behavior when caching reads. When using NVMe and SSD disks, reads will not be cached as this reduces the wear on the capacity drives. Conversely, if SSD and HDD drives are selected, then the reads are also cached on the SSD drives to improve the reads and offer flash-like latency. These two behaviors are shown in the following diagram:

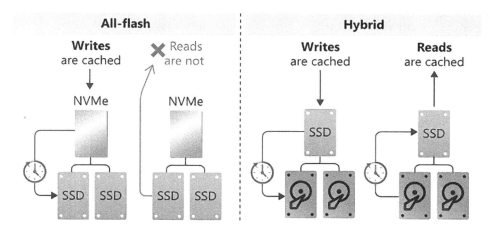

Figure 11.4 – Azure Stack Hub caching behavior

This completes our understanding of drives. Before we look at services, we need to cover volumes.

Volumes

The storage service within Azure Stack Hub will partition the available storage into separate volumes that are used to store both system and tenant data. Volumes will combine the drives in the storage pool to provide the storage space's direct functionality, including scalability, fault tolerance, and performance.

By default, three types of volume types are created in the Azure Stack Hub Storage pool, as follows:

- **Infrastructure volume**: To host files used by Azure Stack Hub infrastructure virtual machines and core services

- **VM temp volume**: To host temporary disks attached to tenant virtual machines and associated data

- **Object store volume**: To host tenant data such as blobs, tables, queues, and virtual machine disks

Fault tolerance is provided by mirroring from within the storage space's direct component of Azure Stack Hub. This is achieved by making multiple copies of the same data. Each copy of the data is written to different physical hardware.

Mirroring in Azure Stack Hub is performed as a three-way mirror that ensures data resiliency. This means that an Azure Stack Hub deployment can tolerate two hardware failures – either drive or server – with no data loss. Three copies of the tenant data are written to different servers via the cache layer, as shown in the following diagram:

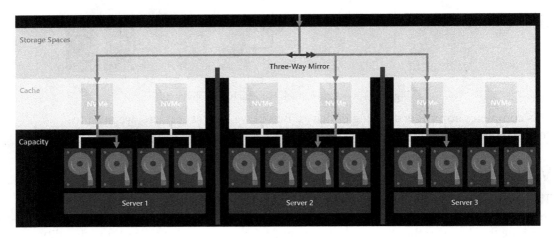

Figure 11.5 – Azure Stack Hub three-way mirroring

This completes our run-through of the infrastructure components that underpin storage in Azure Stack Hub. We will now look at different services provided by Azure Stack Hub Storage, beginning with disks.

Disks

The disk services provided by Azure Stack Hub are similar to those supplied by the Azure public cloud but with a couple of differences, as shown in the following diagram:

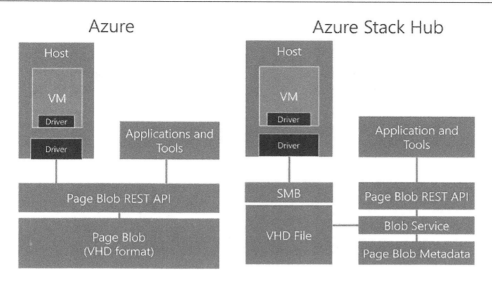

Figure 11.6 – Azure Stack Hub disks

As shown in the preceding diagram, the services that are used for disks are broadly the same as the page blob and REST APIs.

The virtual machine attaches the disk, and the page blob is leased by the virtual machine. The virtual machine can then read and write to the disk. This provides the latency, **input/ output operations per second** (**IOPS**), and throughput needed for the virtual machine.

The REST API allows us to upload disks and images for virtual machines, and also provides a snapshot functionality. Unlike the Azure public cloud, no live or incremental snapshots are available through the REST API.

Virtual machine disks can be containers or one or more data disks. For containers, users can select where containers blobs are placed. The storage service is responsible for selecting the volume with the most available space for new containers. The OS and data disks of virtual machines will be stored as page blobs, and each disk will be placed in a separate container.

This interaction is shown in the following diagram:

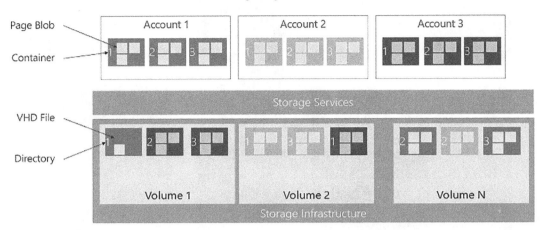

Figure 11.7 – Azure Stack Hub virtual machine disks

Alongside the blobs and containers, the managed disks allow us to easily manage the blobs and storage accounts that are associated with virtual machine disks. Managed disks are enabled in Azure Stack Hub by default when the Azure Stack Hub portal is used to create a virtual machine disk. Managed disks provide disk-level, role-based access and also support locks and tags. Managed disks have a limit of 1 TB within Azure Stack Hub, unlike Azure, which has a 4 TB limit for managed disks. Managed disks in Azure Stack Hub utilize BitLocker for encryption of data at rest. There is currently no migration path between unmanaged and managed disks.

There are a few differences between managed and unmanaged disks. Managed disks do not require a storage account and are allocated as one resource per disk. Unmanaged disks, on the other hand, require the use of a storage account, and one or more disks can be allocated to a storage account. Unmanaged disks can be spread across multiple storage accounts.

Now, let's look at blobs.

Blobs

The blob storage service is used to store and serve unstructured data. Blob storage is effective for storing the following content:

- Application and web-scale data
- Big data from IoT
- Backups or archives
- Photos, video, and blogs

Each blob is organized into a container. A storage account can contain an unlimited number of containers, and each container can contain an unlimited number of blobs until the limit of the storage account is reached.

Three types of blobs are offered by blob storage:

- **Block blobs**: Optimized for streaming
- **Append blobs**: Optimized for append operations
- **Page blobs**: Optimized for IaaS disks

Blob storage is scalable and highly available, as well as cost-effective and durable. Blobs are mutable, and the REST API includes efficient continuation and retry. This allows for parallel, out-of-order uploads for large block blobs.

Now that we have covered blobs, let's focus on tables.

Tables

Azure Stack Hub tables are **NoSQL** key-value stores. They offer a flexible schema and enable rapid development. The use of tables allows you to scale from a few records to millions of records and provide a strong consistency model. Azure Stack Hub tables would typically be used for the following scenarios:

- Shopping carts
- Address books
- Server status
- Device information
- Events

The partition key and row key in the Azure Stack Hub Storage table are both limited to 400 characters each and 800 bytes in size. All partitions of a tenant table are stored in the same database. This limits the scalability of a single table to the limit of the transaction rate of a single table server database.

The remaining component we must touch on with regard to this overview of storage is the queue component.

Queues

Queues in Azure Stack Hub are simple, cost-effective, and durable for large workloads. They can be accessed from anywhere using an authenticated HTTP/HTTPS endpoint. They support messages up to 64 KB in size and can support millions of messages within a queue. The storage account can contain an unlimited number of queues. The only limitation regarding the total number of messages is the capacity of the storage account.

The typical use cases for queues include asynchronous messaging between decoupled application components or managing a backlog of work items and scaling at burst times.

This completes our in-depth overview of the different storage services and components. Now, let's look at how we manage storage within Azure Stack Hub, including container management. We will also look at the typical usage of the available storage services.

Reviewing Azure Stack Hub Storage management and usage

In this section, we are going to focus on how to manage and use the Azure Stack Hub Storage services we reviewed in the previous section. We will begin by looking at container management.

Container management

Some considerations need to be taken into account when managing storage within Azure Stack Hub. We must focus on how to manage capacity from an administration point of view, which brings the health and monitoring capabilities of storage accounts in Azure Stack Hub into play.

The capacity view will alert you when a share is running out of capacity or if there is a problem with the availability of a share. The storage service health will alert you if there are problems with the availability of the data store, or if there are errors with any infrastructure dependencies. Some typical alerts you may see and the remedial actions that can be taken are shown in the following table:

Alert Name	Severity	Component	Alert Remediation
An infrastructure role is unresponsive.	Critical	Storage controller	1. Navigate to the Storage controller infrastructure role and restart the role. 2. If the issue persists, contact support.
Internal data store is unavailable.	Critical	SRP	1. Review the alerts on Network Controller Role blade.
A physical disk has failed.	Warning	Capacity management	1. The physical disk located <Location Info> has failed.
Storage Service role is experiencing moderate levels of errors.	Warning	Storage Services	1. Navigate to the Storage services infrastructure role and restart the role.

There are certain actions that an operator can take to ensure the storage container has enough capacity to run the relevant workload for the tenants. One of these is to reclaim capacity that is being used by a tenant account that has been deleted. This capacity is automatically reclaimed when the data retention period is reached, but by using the admin portal, you can reclaim this space immediately. To reclaim the capacity immediately within the admin portal, navigate to the **Storage accounts** panel and select **Reclaim space** at the top of the panel. This will result in a message appearing at the top of the panel, as shown in the following screenshot:

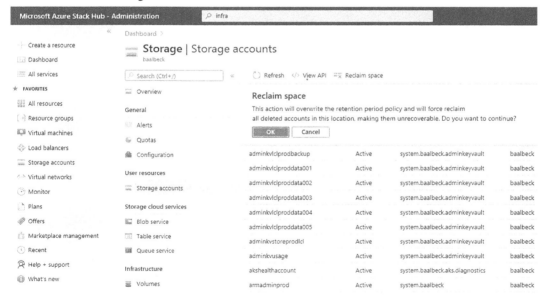

Figure 11.8 – Azure Stack Hub – reclaim storage space

Clicking **OK** will free up the space of the deleted tenant account.

This same procedure can also be performed with PowerShell using the following cmdlets:

```
$farm_name = (Get-AzsStorageFarm)[0].name
Start-AzsReclaimStorageCapacity -FarmName $farm_name
```

It should be noted that this is permanently deletes the tenant data and account, which means they can't be recovered.

Due to the way tenants make use of the platform, some of the tenant shares end up using more space than others. This can result in a share that runs low on space while other shares are relatively unused. Blob container migration does not currently support live migration, so migrating a container between volumes must be performed as an offline operation. To migrate managed disks, you can use the PowerShell cmdlets shown here:

```
$ScaleUnit = (Get-AzsScaleUnit)[0]
$StorageSubSystem = (Get-AzsStorageSubSystem -ScaleUnit
$ScaleUnit.Name)[0]
$Volumes = (Get-AzsVolume -ScaleUnit $ScaleUnit.Name
-StorageSubSytem $StorageSubSystem.Name | Where-Object {$_.
VolumeLabel -Like "ObjStore_*"})
$SourceVolume = ($Volumes | Sort-Object RemainingCapacityGB)[0]
$VolumeName = $SourceVolume.Name.Split("/")[2]
$VolumeName = $VolumeName.Substring($VolumeName.IndexOf(".")+1)
$MigrationSource = "\\SU1FileServer."+$VolumeName+"\
SU1_"+$SourceVolume.VolumeLabel
$Disks = Get-AzsDisk -Status All -SharePath $MigrationSource |
Select-Object -First 10
$DestinationVolume = ($Volumes | Sort-Object
RemainingCapacityGB -Descending)[0]
$VolumeName = $DestinationVolume.Name.Split("/")[2]
$VolumeName = $VolumeName.Substring($VolumeName.IndexOf(".")+1)
$MigrationTarget = \\SU1FileServer."+$VolumeName+"\
SU1_"+$DestinationVolume.VolumeLabel
$jobName = "MigrationDisk"
Start-AzsDiskMigrationJob -Disks $Disks -TargetShare
$MigrationTarget -Job $jobName
```

This works for managed disks only, not for unmanaged disks. To manage space for unmanaged disks, you must distribute the unmanaged disks across multiple volumes. However, since this is a complex task, it should only be conducted under guidance from Microsoft support.

The other consideration that needs to be taken into account when it comes to container management is the retention period, which we briefly mentioned earlier in this section. When configuring storage, you can allocate the number of days that data is retained after a storage account is deleted. This can be set to anything between 0 and 9,999 days. Setting the value to 0 means that the data contained within the storage container will be deleted as soon as the account is deleted. This retention period is defined on a region-by-region basis, and the default retention period is 15 days.

This completes our review of container management. Now, let's take a quick look at the storage panel within the Azure Stack Hub admin portal, which shows the usage of the storage account.

Storage usage

The Azure Stack Hub administration portal provides a comprehensive panel for managing the usage and capacity of storage accounts. An example of the storage panel for a tenant is shown in the following screenshot:

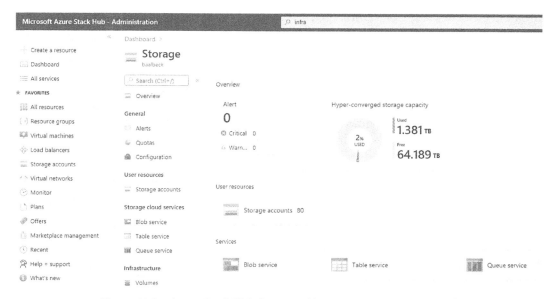

Figure 11.9 – Azure Stack Hub Storage – Tenant storage capacity panel

This panel allows an operator to drill down to see exactly what storage is being consumed across the different containers and storage accounts for a particular tenant.

Finally, we will look at data transfer tools.

Data transfer tools

Azure Stack Hub, through the use of storage services, requires you to manage or move data to or from Azure Stack Hub. To do this, we can use the following Azure storage tools:

- AzCopy: A command-line utility designed specifically for storage and allows you to copy data from one object to another

- Azure PowerShell: A command-line shell designed to help you easily perform system administration tasks

- Azure CLI: An open source tool that provides a defined set of commands for working with Azure Stack Hub

- Azure Storage Explorer: A standalone application that provides a graphical user interface for performing storage tasks

- Blobfuse: A virtual filesystem driver specifically designed for Azure Blob storage that utilizes the Linux operating system

This completes our review of the storage components and services within Azure Stack Hub. Now, let's look at the available compute components.

Overviewing Azure Stack Hub compute

We will begin this section by providing an overview of the Azure Stack Hub compute solution. Azure Stack Hub provides IaaS functionality such as virtual machines, disks, and networking. Virtual machines are the first step for IaaS within Azure Stack Hub and are the touchstone for Azure consistency. Almost every application that runs on Azure Stack Hub relies on some form of IaaS infrastructure, be that middleware, SQL Server, or Exchange Server. All PaaS offerings build off of IaaS, including Service Fabric, notification hubs, and web apps. All of these are supported by foundational services provided by Azure Stack Hub. Azure Stack Hub uses these foundational services to help it run hybrid workloads, which is what Azure Stack Hub is designed for. The following diagram shows an Azure Stack Hub solution:

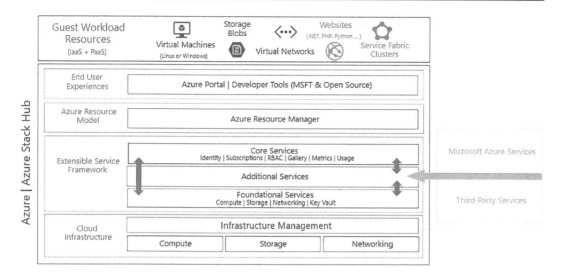

Figure 11.10 – Azure Stack Hub solution view

Azure Stack Hub can be extended by the extensible service framework when additional functionality is added to Azure to allow Azure Stack Hub to keep pace. It can also be extended by introducing third-party services designed to work with Azure Stack Hub.

We covered Azure Resource Manager (ARM) templates and resource providers in *Chapter 8, Working with Offers, Plans, and Quotas*, where we worked with ARM templates and one of the resource providers that can be utilized is the compute resource provider. The ARM template allows us to define the desired state of a deployed resource, such as a Windows virtual machine.

Azure Stack Hub enables several IaaS workloads, and samples can be downloaded from GitHub at `https://github.com/Azure/azurestack-quickstart-templates`.

As we mentioned in *Chapter 8, Working with Offers, Plans, and Quotas*, we can work with ARM templates to specify all the resources needed to make up a solution. This can include one or more virtual machines that are supported by the underlying foundational services, including compute.

Now that we have looked at how the compute feature fits into the Azure Stack Hub environment, let's look at the different components that make up the compute feature within Azure Stack Hub.

Understanding Azure Stack Hub compute components

Let's take a look at the compute components within Azure Stack Hub with virtual machines.

Virtual machines

The compute services within Azure Stack Hub are used to support virtual machines. It supports both Windows and Linux virtual machines, both of which can be created from custom or syndicated images. Azure Stack Hub supports a subset of virtual machine sizes compared to Azure public cloud. The virtual machine types that are supported are as follows:

- Entry-level:

  - A\Basic

  - A\Standard

  - Av2

- General-purpose:

  - D

  - DS

  - DSv2

  - Dv2

  - Dv3

  - Ev3

- Compute-intensive:

  - F

  - FS

  - FSv2

- GPU:

  - NCv3

  - NVv4

  - NcasT4_v3

The performance of the virtual machines depends on the underlying infrastructure, so unlike Azure, there is no guarantee that the same virtual machine will perform the same across multiple Azure Stack Hub deployments.

Virtual machine scale sets (VMSSes) are built into the Azure Stack Hub platform, but there is no insights-based auto-scaling available in the platform.

The definitions of the RAM, CPU, and disk size quantities in Azure Stack Hub are aligned with the same values that are defined in Azure. However, CPU performance is non-deterministic for a given virtual machine size due to the nature of the Azure Stack Hub infrastructure. Each deployment of the Azure Stack Hub infrastructure will potentially be different, and some virtual machine sizes will perform better than others.

The Azure Stack Hub placement engine is responsible for deciding where to place the tenant virtual machines across the available nodes in the Azure Stack Hub environment. Two considerations come into play when this placement decision is being made. Firstly, we must consider the amount of memory that the virtual machine will consume and whether the host will have enough memory available. Secondly, we must consider whether the virtual machine is included in an availability set or included in a VMSS.

Availability sets are used to provide high availability for the compute layer in Azure Stack Hub. An availability set is used to ensure that virtual machines are spread across multiple fault domains. A fault domain, when referenced as part of an availability set, is defined as a single node in a scale unit. Virtual machines that are placed into an availability set will be physically isolated from each other by spreading them across multiple fault domains. A four-node Azure Stack Hub environment can handle an availability set with three fault domains. One node is reserved for use during patch and update activities.

This is shown in the following diagram:

Figure 11.11 – Azure Stack Hub fault domains

In the previous diagram, each host is colored so that they match a fault domain for a particular availability set.

Let's continue with this same four-node Azure Stack Hub. The following diagram shows an example of two different availability sets spread across the four nodes:

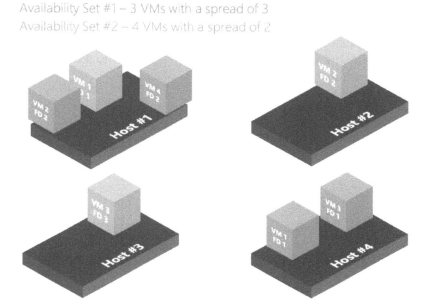

Figure 11.12 – Azure Stack Hub availability sets

VMSSes rely on availability sets to make sure each VMSS is distributed to a different fault domain. This means that the VMSSes should be spread across the available nodes in the Azure Stack Hub environment. This distribution is designed to allow the Azure Stack Hub environment to cope with a hardware failure and still keep its workload running.

Since the 1901 update of Azure Stack Hub, there is now a limit on the total number of virtual machines that can be created in an Azure Stack Hub environment. The limit has been set to 60 virtual machines per server and a total of 700 across the full Azure Stack Hub environment. In the previous example of a four-node system, the limit would be 240 total virtual machines, 60 per node * 4 nodes.

The ability of Azure Stack Hub to be able to continue running workloads in the event of a failure is based on the available memory. Allowing Azure Stack Hub to manage this relies on reserved memory, which is not available for tenant workloads. This can be monitored within the Azure Stack Hub administration portal for each scale unit, as shown in the following screenshot:

Figure 11.13 – Azure Stack Hub physical memory

Several components go into the used memory segment, including the following:

- **Host OS usage:** The memory that is used by the OS on the host, virtual memory page tables, and the storage space's direct memory cache

- **Infrastructure services**: Infrastructure virtual machines for Azure Stack Hub

- **Resiliency reserve**: Portion of memory reserved to allow tenant availability

- **Tenant virtual machines**: Tenant workload running on Azure Stack Hub

- **Value add resource provider**: Any additional resource provider that's been installed, including event hubs

These components affect what memory is available for placing the new workload in a given node in the Azure Stack Hub environment.

Before we move on and look at the next component of compute resource providers, we will cover VMSSes, which we mentioned earlier in this section.

Virtual machine scale sets

Azure Stack Hub includes a compute resource known as VMSSes. These can be used to deploy and manage a defined set of identical virtual machines. All of the virtual machines that are defined in the VMSS are configured in the same way, so the virtual machines don't need to be pre-provisioned. As VMSSes do not support autoscaling, you will need to add additional instances to the VMSS using ARM templates, the Azure CLI, or PowerShell.

Before we can create a VMSS in Azure Stack Hub, the Azure Stack Hub must be registered with Azure to ensure the availability of marketplace items. The virtual machine images to be used for the VMSS must be downloaded from the marketplace before they are created. A scale set can be created using the Azure Stack Hub portal in the same manner that virtual machines are created.

Now, let's look at the compute resource provider.

Compute resource provider

The compute controller is an infrastructure component that manages compute clusters and orchestrates virtual machine creation. This was born from Azure compute and leverages code that has been hardened in Azure. The compute architecture is based on the microservice architecture, with resiliency provided by running all the services in a service fabric ring. This compute architecture is shown in the following diagram:

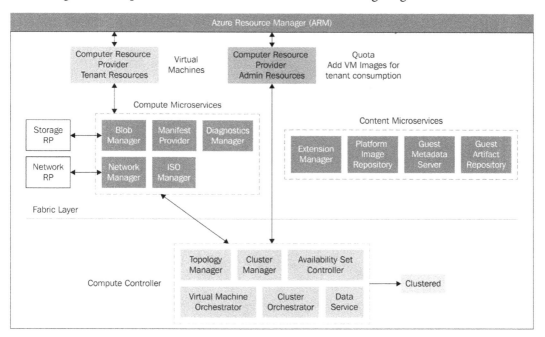

Figure 11.14 – Azure Stack Hub compute architecture

This completes this section on the compute components within Azure Stack Hub. Now, let's review the compute scenarios from an operator viewpoint in Azure Stack Hub.

Reviewing Azure Stack Hub compute scenarios

During deployment, the infrastructure virtual machines are registered to the compute controller or **compute resource provider** (**CRP**). These infrastructure virtual machines are created as Gen2 services with dynamic memory, while tenant virtual machines are created as Gen1 services. One particular infrastructure virtual machine in the portal is an infrastructure role instance. These virtual machines can be started, restarted, or shut down via the Azure Stack Hub portal. CRP uses placement rules to prevent all the infrastructure virtual machines of the same type from running on the same host. An example of this infrastructure instance in the Azure Stack Hub administration portal is shown in the following screenshot:

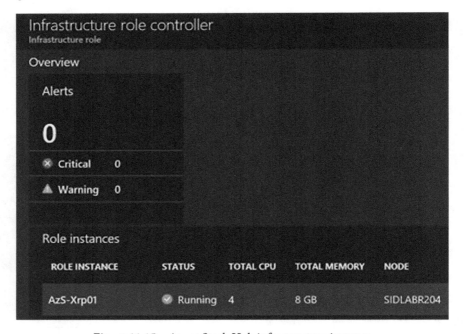

Figure 11.15 – Azure Stack Hub infrastructure instance

Various life cycle operations can be performed by the compute controller and kicked off from the Azure Stack Hub portal, as shown in the following screenshot:

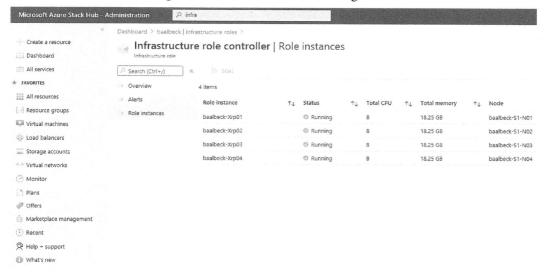

Figure 11.16 – Azure Stack Hub infrastructure instance life cycle operations

Let's walk through the process of creating a virtual machine by using a creation pipeline.

Virtual machine creation

The compute resource provider's virtual machine pipeline is a goal-seeking engine that exercises the full breadth of the core services in Azure Stack Hub to create and place virtual machines. This pipeline is shown in the following diagram:

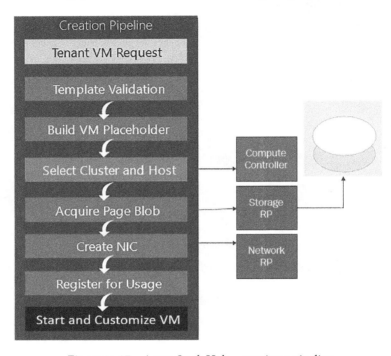

Figure 11.17 – Azure Stack Hub – creating a pipeline

CRP tenants can bring their own images or leverage images that have been added to the CRP's virtual machine image gallery.

An Azure Stack Hub operator can track the consumption of the CRP's services and the tenant's usage.

The compute controller uses Windows Server compute features, including failover clustering and Hyper-V, as shown in the following diagram:

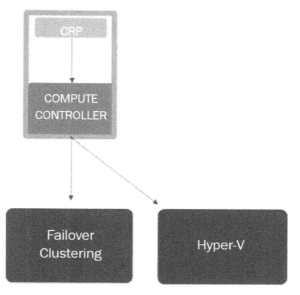

Figure 11.18 – Azure Stack Hub CRP

Now that we've looked at the virtual machine life cycle, let's look at the cluster life cycle.

Cluster compute life cycle

The scale unit and its associated nodes are registered against the compute controller during deployment. The compute controller drains nodes for field replacement unit scenarios and is also responsible for orchestrating the power cycle of scale unit nodes. Virtual machine clustering provides high availability. The movement of virtual machines maintains high availability, even during server down conditions such as patch and update cycles. Since extra nodes are added during scale-out operations, these will, in turn, be registered against the compute controller.

With that, we've looked at the compute features within Azure Stack Hub. Before we move on to the next chapter, let's quickly recap on what we have learned in this chapter.

Summary

In this chapter, we looked at two of the underlying core services of Azure Stack Hub in terms of storage and compute. We began by providing an overview of the storage feature in Azure Stack Hub and the different components that make up this feature. This introduced us to the components that underpin Azure Stack Hub Storage, including tables, queues, blobs, and disks, among others. From there, we looked at storage management and usage. This introduced us to processes such as reclaiming storage when accounts are deleted and using alerts to control how storage accounts are used to prevent workload failure. After this, we reviewed the compute component within Azure Stack Hub.

We built on top of *Chapter 8, Working with Offers, Plans, and Quotas*, by working with ARM templates and provided a walkthrough of the virtual machine creation process. We were introduced to the compute controller, so we should now have an understanding of how this fits into the overarching architecture of Azure Stack Hub. Now that we've looked at these features, in the next chapter, we will focus on how to monitor and manage Azure Stack Hub.

Section 4: Monitoring, Licensing, and Billing

The final part of this book covers the remaining components of Microsoft Azure Stack Hub, which are included as part of the Microsoft AZ-600 exam for monitoring, licensing, billing, and support.

The following chapters will be covered under this section:

12
Monitoring and Managing Azure Stack Hub

This chapter is dedicated to managing and monitoring Microsoft Azure Stack Hub and covers not only the components within Microsoft Azure Stack Hub, but also integrating them with existing tools that customers may already be utilizing. We will touch on alerting, capacity, and resource providers. We will also cover Microsoft Azure Monitor and how it can be used within Microsoft Azure Stack Hub.

In this chapter, we will cover the following topics:

- Understanding the basics of Azure Stack Hub management and monitoring
- Reviewing Azure Stack Hub monitoring
- Integrating Azure Stack Hub monitoring with existing tools
- Managing Azure Stack Hub

We will begin by understanding the basics of Azure Stack Hub management and monitoring.

Technical requirements

You can view this chapter's code in action here: `https://bit.ly/3zbQhvU`

Understanding the basics of Azure Stack Hub management and monitoring

Azure Stack Hub software components are internally monitored via the health service and the Service Fabric health service. Metrics and logs are centrally collected. **Original equipment manufacturer (OEM)** vendors are responsible for agentless hardware device monitoring through external systems. This is achieved by each OEM, adding a host to Azure Stack Hub called the **hardware life cycle host (HLH)**. Storage Spaces Direct storage is monitored by Azure Stack Hub components. A REST API is available for the third-party integration of alerts, metrics, and logs. This includes the **System Center Operations Manager (SCOM)** Azure Stack Hub Fabric management pack and Nagios integration. This infrastructure is shown in the following diagram:

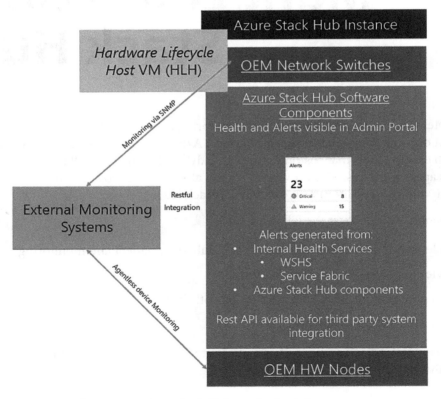

Figure 12.1 – Azure Stack Hub monitoring infrastructure

Infrastructure resource providers enable communication with the underlying infrastructure roles and hardware layer. Microservices running in the infrastructure service ring are supported by the infrastructure resource providers. These infrastructure resource providers include **fabric resource providers** (**FRPs**), **health resource providers** (**HRPs**), and **updates resource providers** (**URPs**). These infrastructure resource providers can only be called with the admin subscription. Each infrastructure resource provider provides a northbound REST API. These REST APIs can be consumed via the Azure portal, PowerShell, Visual Studio, and other tools. Each infrastructure resource provider uses a southbound API to communicate with the controllers. The health and update resource providers also provide a registration interface for **ARM** deployable resource providers. These layers are shown in the following diagram:

Figure 12.2 – Azure Stack Hub infrastructure resource providers

As an Azure Stack Hub administrator, you can use the portal to ask the fabric resource provider to perform an action. The action will be performed by the appropriate Azure Stack Hub service. This includes operations such as changing the boot order, power on, and power off, as shown in the following diagram:

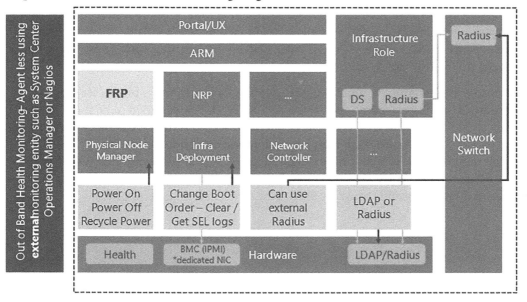

Figure 12.3 – Azure Stack Hub hardware monitoring

These are the basics of the infrastructure that underpins the monitoring capabilities within Azure Stack Hub. Now that we have an understanding of these basics, we can review Azure Stack Hub monitoring.

Reviewing Azure Stack Hub monitoring

Before we look at monitoring in Azure Stack Hub, let's look at the monitoring problem space.

First, understanding health is complex for Azure Stack Hub. A cloud architecture such as Azure Stack Hub depends on many different technologies. The health of the Azure Stack Hub cloud is not the sum of its parts. This means that we need to understand how alerts relate to health.

Second, we need to be able to decipher the alerts to understand which alerts need to have actions taken against them. We need to prioritize which alerts to tackle first and also need to understand the steps needed to resolve the alerts.

The philosophy of monitoring for Azure Stack Hub is to show the health of the platform through alerts. These alerts are managed by the **health resource provider (HRP)**. You can consume these alerts to suit your own tools and processes. Azure Stack Hub manages and monitors all the software in the box. The OEM vendor provides the management capabilities for the hardware, including in-box network switches. As an example, Lenovo provides a utility called **XClarity** that includes controllers in each of the components, an administration portal, and an integrator into the Azure Stack Hub administration portal. The OEM-specific virtual machine that's within the HLH hosts these hardware management roles, and it can also be the target location to store logs that have been pulled from an Azure Stack Hub scale unit during troubleshooting.

Now that we understand the problem space, we can look at the core principles of how Azure Stack Hub monitors and operates. Azure Stack Hub is designed to be treated as an appliance. The majority of a cloud admin's interaction with Azure Stack Hub should be driven by alerts. The Azure Stack Hub health component should be directly tied to alerts. These alerts should make it easy to understand the impact and relevant next steps. They should be able to be resolved using common Azure Stack Hub admin action patterns. They should also be specific to Azure Stack Hub.

The following are some alerts and remediation examples:

Alert Name	Severity	Component	Alert Remediation
An infrastructure role is unresponsive	Critical	Compute Controller	Navigate to the compute controller infrastructure role and restart the role. If the issue persists, contact support.
Low memory capacity in Azure Stack Hub region	Warning	Capacity Management	Add a node to the scale unit using the Capacity management admin blade.
A physical disk has failed	Warning	Capacity Management	The physical disk located at <Location Info> has failed. The process of repairing the storage virtual disk has been initiated. It is important to replace the physical disk to ensure full capacity and resiliency. Click here to learn more about performing the disk replacement procedure.

Now that we understand the core principles, we can look at the tools and options for monitoring Azure Stack Hub, starting with Azure Monitor.

Azure Monitor on Azure Stack Hub

Azure Monitor is a platform service that provides a single source for monitoring all Azure resources. Through the use of Azure Monitor, you can visualize, query, route, archive, or otherwise take action on the metrics and logs coming from the resources in Azure Stack Hub and have the same experience as using Azure. You can work with this data by using the Azure Stack Hub admin portal, PowerShell cmdlets, a cross-platform CLI, or Azure Monitor REST APIs. Azure Monitor on Azure Stack Hub provides metrics for the compute and storage services, with additional services to be added in the future. The following screenshot shows Azure Monitor in Azure Stack Hub:

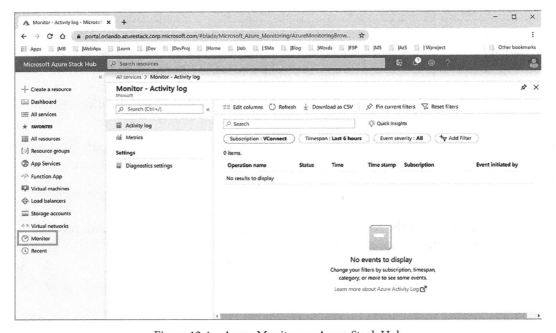

Figure 12.4 – Azure Monitor on Azure Stack Hub

Applications can run in the operating system of a virtual machine running with the `Microsoft.Compute` resource provider. These applications and virtual machines emit their own sets of logs and metrics. Azure Monitor makes use of the Azure diagnostics extensions (both Windows and Linux) to collect most application-level metrics and logs. Azure Monitor can provide feedback on the following types of measures:

- Performance counters
- Application logs
- Windows event logs

- .NET event sources

- IIS logs

- Manifest-based ETWs

- Crash dumps

- Customer error logs

The `Microsoft.Compute` resource provider covers both virtual machines and virtual machine scale sets. The following diagram shows the compute subset for Azure Monitor:

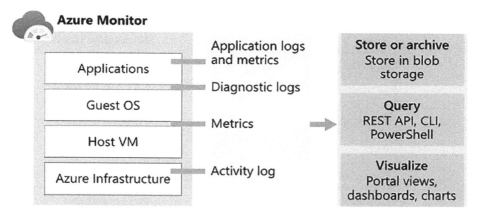

Figure 12.5 – Azure Monitor compute subset

Metrics that have been captured in Azure Monitor are stored for 90 days. Activity logs that have been captured in Azure Monitor are also stored for 90 days. Not all monitoring data is stored in Azure Monitor, however, such as diagnostic logs. To increase the retention period of this collected data, it can be archived in a storage account.

Azure Monitor provides different tools, including Azure Monitor REST API, cross-platform **command-line interface (CLI)** commands, PowerShell cmdlets, and the .NET SDK to access and query the data in the system or storage account.

Azure Monitor can help you visualize your monitoring data in graphics and charts that help you find trends quicker than trying to look through the data itself. Both the Azure Stack Hub user and administration portals have visualization built into them, which means they can interact with the Azure Monitor data. It is also possible to route this data from Azure Monitor to Microsoft Power BI or third-party visualization tools.

The metrics and diagnostic logs that can be captured by Azure Monitor beyond the `Microsoft.Compute` resource provider vary, depending on the resource type.

The following table lists some of the metrics that are available within Azure Monitor's metric pipeline in Azure Stack Hub:

Resource Provider	Metric	Unit	Aggregation Type	Description
virtualMachines	Percentage CPU	Percent	Average	Virtual machine CPU allocated percentage
storageAccounts	Used Capacity	Bytes	Average	Account used capacity
	Transactions	Count	Total	Storage requests
	Ingress	Bytes	Total	Input data in bytes
	Egress	Bytes	Total	Output data in bytes
	Availability	Percent	Average	Average availability measurement as a percentage
blobServices	Blob Capacity	Bytes	Total	How much of the storage account has been used in bytes
	Blob Count	Count	Total	Total number of blobs
	Container Count	Count	Average	Average number of containers
	Transactions	Count	Total	Total requests
	Ingress	Bytes	Total	Inbound data
	Egress	Bytes	Total	Outbound data
	Availability	Percent	Average	Availability as a percent
tableServices	Table Capacity	Bytes	Average	Average storage amount
	Table Count	Count	Average	Average number of tables
	Table Entity Count	Count	Average	Average number of table entities
	Transactions	Count	Total	Total requests
	Ingress	Bytes	Total	Inbound data
	Egress	Bytes	Total	Outbound data
	Availability	Percent	Average	Availability of service

Resource Provider	Metric	Unit	Aggregation Type	Description
queueServices	Queue Capacity	Bytes	Average	Average queue storage
	Queue Count	Count	Average	Average number of queues
	Queue Message Count	Count	Average	Average queue messages
	Transactions	Count	Total	Total requests
	Ingress	Bytes	Total	Inbound data
	Egress	Bytes	Total	Outbound data
	Availability	Percent	Average	Availability as a percentage

> **Note**
>
> Azure Monitor relies on the `Microsoft.Insights` resource provider, so you must ensure this has been registered in the subscription's Offer resource provider settings.

With that, we've looked at how to use Azure Monitor in Azure Stack Hub. In the next section, we will look at the health resource provider and how it works.

Exploring the Health resource provider

The health resource provider is a resource provider that provides and allows actions to be performed on Azure Stack Hub alerts. The health resource provider also collects and displays metric data. It is run as a Service Fabric service in the **core resource provider** (**XRP**) virtual machine. The health resource provider provides a list of resource providers and infrastructure roles to the portal UI, which allows an Azure Stack Hub administrator to assess the health of Azure Stack Hub elements.

The data flow of the health resource provider is shown in the following diagram:

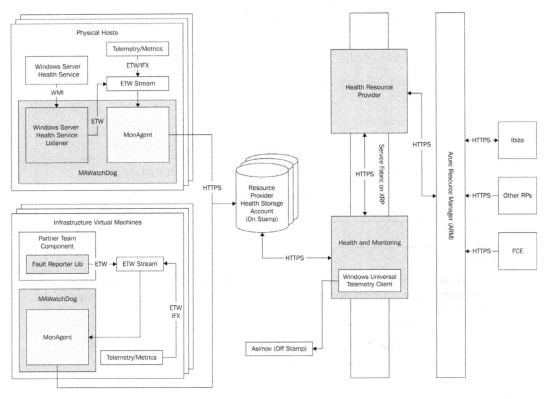

Figure 12.6 – Health resource provider data flow

Raw data is collected from hosts and virtual machines, including faults, heartbeats, metric data, telemetry events, and activity logs. The monitoring system accesses the raw data to process faults that it then converts into alerts, perform metric aggregation, and delete old data. The telemetry subsystem extracts telemetry events and sends them to Microsoft through the standard telemetry interface. This data is then analyzed by Microsoft to understand stamp behavior. Sending telemetry data can be disabled via Azure Stack Hub. Alerts are uploaded and made visible in the portal so that the administrator can respond to them.

All foundational resource providers are registered with the health resource provider during deployment. Additional value-added resource providers such as MySQL do not currently register with the health resource provider. Each registration with the health resource provider supplies the resource provider's display name, storage account details for monitoring, and alert templates. Infrastructure resources are also registered and linked to the health resource provider. This registration data is used to determine which resource providers or resources are healthy and which require attention. The alert templates determine which alerts are shown to the user, and each resource provider defines its own alert templates.

The Azure Stack Hub health and monitoring subsystem can also raise alerts against hardware components, including the following items:

- System fans
- System temperature
- Power supply
- CPU
- Memory
- Boot drives

You would need to engage with your OEM vendor to ensure that this feature is available on their specific hardware and get guidance on how to enable it.

With that, we now understand the health resource provider and how it is used to generate alerts. Now, let's dive deeper into these alerts and look at some of the remediation operations we can perform.

Alert remediation

As we mentioned at the beginning of this chapter, administrative actions that are performed by an operator or administrator are driven by alerts. This is driven by a resource action hierarchy that dictates how actions are applied when recovering functionality. For example, on an infrastructure role instance, you can perform the start, shutdown, and restart operations, while for a scale unit node resource, you can also perform the repair, drain, resume, power off, and power on operations.

Let's look at an example where an infrastructure role instance such as Hyper-V is unresponsive, and its functionality needs to be restored. In this instance, we can only use the start, shutdown, and restart actions.

Some alerts provide support for a repair option. If this is available, then when the repair action is selected, it performs steps to attempt to resolve the issue using steps that are specific to the alert. This repair action will report either a success or a failure once all the steps have been completed. The alert will automatically close once the repair has been successfully completed.

Not all alerts provide this repair option, so they must be manually resolved. The alert will detail the steps that need to be performed to resolve the alert. You must then manually close the alert. If the steps have not been completed correctly, then a new alert will be generated.

Any alerts that are raised on hardware such as boot drives will not automatically close once the defective part has been replaced; they must be closed manually.

Now that we have an idea of what alert remediation is, we have finished looking at the internal monitoring provided by Azure Stack Hub. This is only one part of the solution; what if you need to connect to external monitoring/capacity/ticketing systems? We will cover this in the next section, where we will look at integrating with existing tools.

Integrating Azure Stack Hub monitoring with existing tools

To be able to query the Azure Stack Hub's state and health, you must connect to sufficient authorization and use an API to talk to the fabric, health, or update resource providers. Each of these API calls will follow a similar pattern, as shown in the following diagram:

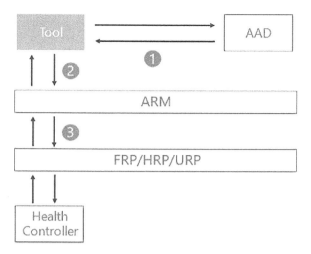

Figure 12.7 – Azure Stack Hub Health API

In the preceding diagram, the flow will run as follows:

1. Authenticate with **AAD.**

2. Retrieve a token for an admin subscription.

3. Call the API namespace with an action.

More details about the REST API can be found at `https://docs.microsoft.com/en-us/rest/api/azure-stack/`.

Using the fabric resource provider will get you access to the following namespace resources:

- Location (region)
- Infrastructure roles
- Logical networks
- Storage subsystem
- File shares
- Role instances
- Scale units
- Physical servers

For each of these namespace resources, you can perform the `GET`, `PUT`, and `POST` actions.

Using the health resource provider will give you access to the following namespace resources:

- Location (region)
- Alert
- Service health registrations
- Resource health registrations

For each of these namespace resources, you can perform the GET, PUT, and DELETE actions.

Take a look at the following sample commands, which will query Azure Stack Hub for current alert properties:

```
# // Variables


# Enter your tenant name or id below
$tenantid = "yourdomain.onmicrosoft.com"
# Enter credentials with permissions to the Azure Stack
$username = 'admin@yourdomain.onmicrosoft.com'
$password = '555-hack-me-please'


// Main Routine

# Get a bearar token
# Note: The client_id, grant_type and scope values are
constants
# Note: Use the $bearerTokenResource from the sample above
$body = @{
    grant_type = "password"
    client_id = "1950a258-227b-4e31-a9cf-717495945fc2"
    resource = $bearerTokenResource
    username = $username
    password = $password
    scope= "openid"
}


# Store token
$method = "POST"
```

```
$uri = "$tenantid/oauth2/token"
$token = Invoke-RestMethod $uri -Body $body -Method $method
-ErrorAction Stop -ContentType 'application/x-www-form-
urlencoded'
# Output the token
$token
# The API version of the endpoint
$apiversion = "2015-11-01"
# Enter the region for your stamp below
$region = "north"
# Enter the dns for your stamp below
$dns = "my.azurestack"
# Enter the subscription id for your default provider
subscription
$subscriptionid = "a44de15f-f210-4578-840c-c311d962d3ef"
# Enter the resource group name where you store tenant azure ad
registrations
$resourcegroup = "custom.registrations"


# // Main Routine


# Setup the headers for the request
# Note: use the $token variable for the sample above
$headers = @{}
$headers.Add("Authorization","$($token.token_type) "+ " " +
"$($token.access_token)")



$URI="${ArmEndpoint}/subscriptions/${subscription}/
resourceGroups/system/providers/Microsoft.
InfrastructureInsights.Admin/regionHealths/$region/Alerts?api-
version=2016-05-01"
$Alert=Invoke-RestMethod -Method GET -Uri $uri -ContentType
'application/json' -Headers $Headers
$Alerts=$Alert.value
$Alertsprop=$alerts.properties
$Alertsprop |select
alertid,state,title,resourcename,createdtimestamp,remediation
```

The last resource provider is the update resource provider, which will give you access to the following namespace resources:

- Location (region)
- Region update status
- Updates:

  - Update runs

For each of these namespace resources, you can only use the GET and PUT actions.

You can enable IT Service Management by adopting existing pipelines, which will allow you to use existing connections from monitoring to ticketing and others, as shown in the following diagram:

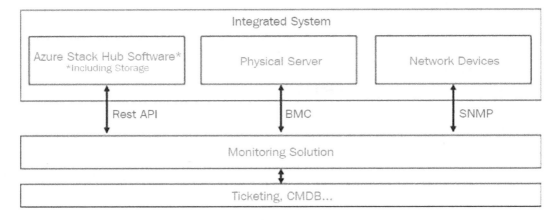

Figure 12.8 – Azure Stack Hub ITSM integration

For each of the pipelines in the preceding diagram, there are monitoring integration solutions available for Azure Stack Hub.

The Azure Stack Hub Software REST API is utilized by the Nagios plugin known as SCOM – Azure Stack Management Pack. There are also examples and documentation for custom integration regarding the API on GitHub.

The SNMP pipeline for network devices is used by the Nagios switch plugin and SCOM – Network Device Discovery. This is also available to monitoring solutions provided by the OEM vendor, such as XClarity from Lenovo.

The physical server's BMC is utilized in the Nagios plugins from the OEM vendor, as well as the SCOM – OEM Vendor management pack, if it's available.

Finally, tenant subscription health monitoring is enabled through the use of SCOM – Azure Management Pack or the **Operations Management Suite (OMS)**.

The SCOM – Azure Stack Hub management pack provides the following capabilities:

- Manages multiple Azure Stack Hub deployments
- Supports both AAD and ADFS
- Retrieves and closes alerts
- Integrates the health and capacity dashboard
- Auto-maintains mode for patch and update processes
- Force updates tasks for deployment and region
- Supports notifications and reporting

You can also integrate operations manager with System Center service manager as an end-to-end ticketing solution. This would provide bi-directional communication, allowing you to close an alert in Azure Stack Hub and operations manager once a service request in the service manager has been resolved. This integration is shown in the following diagram:

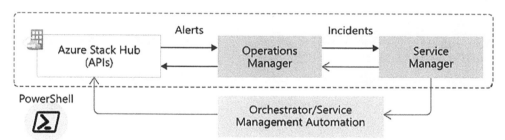

Figure 12.9 – Azure Stack Hub operations manager integration

Alongside this integration with Microsoft tools, Microsoft has also partnered with CloudBase Solutions to develop a Nagios monitoring plugin that is available using the permissive free software license.

This is a Python-based plugin that utilizes the health resource provider REST API, as described earlier in this section. It can retrieve and close alerts in Azure Stack Hub. As with the previous integration with System Center, you can add multiple Azure Stack Hub deployments and send notifications.

The latest version of the Nagios plugin uses the Microsoft Azure **Active Directory Authentication Library** (ADAL), so it supports authentication via a service principal with a secret or certificate. This also allows the plugin to support both **Azure Active Directory** (AAD) and **Active Directory Federation Services** (ADFS) for identity management based on the Azure Stack Hub deployment.

The Nagios version that is going to be integrated through the use of this plugin must be running version 4.x and above. The Microsoft AAD Python library must also be installed. This library can be installed by using the Python `pip` command, as shown in the following Bash script code sample:

```
$ sudo pip install adal pyyaml six
```

Once you have downloaded the plugin from `https://exchange.nagios.org/directory/Plugins/Cloud/Monitoring-AzureStack-Alerts/details`, copy the plugin `azurestack_plugin.py` file to the `/usr/local/nagios/libexec` directory on the Nagios server. The `azurestack_handler.sh` file should be copied to the `/usr/local/nagios/eventhandlers` directory.

Once these files have been copied across, we need to ensure that the plugin has been set to be executed using the following Bash commands:

```
$ sudo cp azurestack_plugin.py <PLUGINS_DIR>
$ sudo chmod +x <PLUGINS_DIR>/azurestack_plugin.py
```

The remainder of the files from the download are configuration files. These should be modified with the relevant information for the Azure Stack Hub deployment being integrated to.

Once they have been updated, these files should be copied to `/usr/local/nagios/etc/objects`.

Then, the Nagios configuration needs to be updated so that the Azure Stack Hub plugin is loaded. This can be done by editing the `/usr/local/nagios/etc/nagios.cfg` file in Bash.

Add the following lines to the config file:

```
cfg_file=/usr/local/nagios/etc/objects/azurestack_contacts.cfg
cfg_file=/usr/local/nagios/etc/objects/azurestack_commands.cfg
cfg_file=/usr/local/nagios/etc/objects/azurestack_hosts.cfg
cfg_file=/usr/local/nagios/etc/objects/azurestack_services.cfg
```

Once the configuration file has been updated, Nagios can be restarted using the following command:

```
$ sudo service nagios reload
```

If Nagios is not available and the operations manager is not in use in your environment, then PowerShell can be used to integrate other monitoring tools with Azure Stack Hub. Commands such as `Get-AzsAlert` and `Close-AzsAlert` can be called to manage alerts from outside Azure Stack Hub.

Now, let's look at managing Azure Stack Hub.

Managing Azure Stack Hub

The main tool that's used for managing Azure Stack Hub is the administrator portal. It is also the easiest way to learn the fundamentals of managing the Azure Stack Hub. You can also utilize PowerShell to perform most management tasks, but this will require additional steps prior to performing any task.

The main responsibility when managing Azure Stack Hub is to make services available for your users and ensure they continue to perform well. Other tasks need to be performed by the operator to manage Azure Stack Hub, including the following:

- User accounts
- Role-based access control
- Managing network and storage
- Replace hardware
- Manage updates

We have covered most of these topics throughout this book, so in this section, we will learn how to manage updates using the administration portal.

Managing Azure Stack Hub updates

When working with an integrated system such as Azure Stack Hub, there are essentially four types of update packages. These update package types are as follows:

- Azure Stack Hub software updates
- Azure Stack Hub hotfixes
- Windows Defender definitions
- OEM-provided updates

The Azure Stack Hub software updates are provided by Microsoft, who are responsible for the end-to-end life cycle of these software updates. These can include Windows Server security updates, non-security updates, plus Azure Stack Hub feature updates. These updates can be downloaded directly from Microsoft and will either be full or express updates. The full update updates the host operating system and will require a lengthy maintenance window to complete, while the express update is scoped so that does not update the underlying host operating system. Microsoft generally release multiple software updates per year. You are alerted through the administration portal when a new software package is available from Microsoft when Azure Stack Hub is running in a connected scenario. If Azure Stack Hub has been deployed in the disconnected scenario, then you can subscribe to the Microsoft RSS feed to be notified of any pending updates.

Hotfixes are also provided by Microsoft and are generally designed for a specific issue that may be security-related or time-sensitive. As with software updates, these can be downloaded directly from Microsoft. Typically, hotfixes are cumulative and can be applied quicker, so they require less downtime in a maintenance window. Before you upgrade a major release of the software update, it is important to ensure that all the hotfixes for the current release have been applied. Any hotfixes that are available for the new major version will be applied when the software package update is applied. Hotfixes can be released multiple times a year, depending on the time-sensitivity of the changes.

Unlike the other updates, OEM-provided updates are the responsibility of the hardware vendor who provided the integrated system. These updates can include hardware-related firmware updates, in addition to driver updates and any OEM vendor-supplied software updates. The OEM vendor is also responsible for updating the HLH server and the network switches where applicable. These updates are generally downloaded directly from the OEM vendor. The OEM vendors release multiple updates a year, generally to coincide with the Microsoft releases, but there may be instances when they are released separately. Each OEM vendor will have its own method for advising customers of the availability of any updates.

The update process consists of three steps:

1. Plan the update
2. Upload and prepare the update package
3. Apply the update

To plan for the update, it is important to notify users of any potential service outages while the update is being processed. A suitable maintenance window should be planned for based on the size and severity of the update to be applied.

If Azure Stack Hub is running in the connected scenario, then the updates are automatically imported into the Azure Stack Hub system. If Azure Stack Hub is running in a disconnected scenario, then the update package must be imported into Azure Stack Hub storage using the Azure Stack Hub portal. All updates must be stored in the `updateadminaccount` storage account for them to be visible in the Azure Stack Hub portal. Ideally, you should create a container in the blob storage of this storage account for each Microsoft update.

All OEM vendor updates must be uploaded to Azure Stack Hub storage, regardless of which scenario the Azure Stack Hub instance is running in, be that connected or disconnected. Again, as with the Microsoft updates, these should be uploaded to a container using the blob storage of the `updateadminaccount` storage account.

The updates can then all be applied through the Azure Stack Hub portal using the **Update** blade.

Updating and managing Azure Stack Hub should be conducted through the Azure Stack Hub portal, but for some tasks, there is a **privileged endpoint** (**PEP**), a pre-configured remote PowerShell console that can be used. We will look at how this is managed and used next.

Privileged endpoint

The PEP exposes a limited set of cmdlets with just the right amount of capabilities to be able to perform a task. To access the PEP and make use of these restricted cmdlets, you don't require an admin account but rather a low privileged account. Full transparency and auditing are included in the PEP with every task that is performed via the PEP that's been logged.

The PEP can be accessed via a remote PowerShell session on the virtual machine where the PEP is hosted. In an integrated system provided by an OEM vendor, there will be three instances of these virtual machines (Prefix-ECRS01, Prefix-ECRS02, and Prefix-ECRS03) with each hosted on a different node for resiliency.

On an integrated system, the following command should be run from an elevated PowerShell session to add the PEP as a trusted host to a hardened virtual machine running on the HLH:

```
Set-Item WSMan:\localhost\Client\TrustedHosts -Value '<IP
Address of Privileged Endpoint>' -Concatenate
```

A remote session with the PEP can now be established by using the following command in the same elevated PowerShell session:

```
$cred = Get-Credential

$pep = New-PSSession -ComputerName <IP_address_of_ERCS>
-ConfigurationName PrivilegedEndpoint -Credential $cred
-SessionOption (New-PSSessionOption -Culture en-US -UICulture
en-US)
Enter-PSSession $pep
```

Now, you can access the restricted set of cmdlets. Since all the tasks that are performed via this endpoint are logged, the session must be closed as soon as the tasks have been completed. This can be done using the following command:

```
Close-PrivilegedEndpoint -TranscriptsPathDestination "\\
fileshareIP\SharedFolder" -Credential Get-Credential
```

It is also possible to import this remote PowerShell session into your local machine by using the following commands from an elevated PowerShell session:

```
winrm s winrm/config/client '@{TrustedHosts="<IP Address of
Privileged Endpoint>"}'
$cred = Get-Credential

$session = New-PSSession -ComputerName <IP_address_of_ERCS> '
    -ConfigurationName PrivilegedEndpoint -Credential $cred '
    -SessionOption (New-PSSessionOption -Culture en-US
-UICulture en-US)
Import-PSSession $session
```

With that, we have covered some of the management qualities of Azure Stack Hub through both the portal and PEP. Additional details and management tasks can be reviewed in the Microsoft docs at the following URL: https://docs.microsoft.com/en-us/azure-stack/operator/azure-stack-manage-basics.

With that, we've learned how to manage Azure Stack Hub and completed this chapter. Before we move on, let's quickly recap what we have learned.

Summary

In this chapter, we looked at working with monitoring and managing in Azure Stack Hub. We started with some of the basics of the monitoring subsystem in Azure Stack Hub, which included an introduction to infrastructure resource providers, such as the health resource provider. We went through the monitoring philosophy of Azure Stack Hub, where Azure Stack Hub and how it is managed is controlled by alerts. We introduced the Azure Monitor platform, which provides metrics for compute and storage services in Azure Stack Hub. We walked through how the health resource provider works and how to deal with alerts. We then went through how to integrate monitoring in Azure Stack Hub into other data center tools such as System Center and Nagios. This introduced us to the REST API and the PowerShell commands that can be used from outside Azure Stack Hub to glean information about the health and status of Azure Stack Hub. Finally, we looked at managing Azure Stack Hub, including performing updates and making use of the PEP.

In the next chapter, we will look at the licensing models in Azure Stack Hub, including how to purchase them and some considerations we should make regarding CSP vendors.

13

Licensing Models in Azure Stack Hub

In this chapter, we will walk through the different approaches to licensing when it comes to Microsoft Azure Stack Hub. We will look at the different Microsoft Azure Stack Hub licensing models and walk through how to purchase Microsoft Azure Stack Hub. Finally, we will look at some considerations that need to be taken into account when it comes to **cloud service providers** (**CSPs**).

In this chapter, we will cover the following topics:

- Understanding Azure Stack Hub licensing models
- Overviewing how to purchase Azure Stack Hub
- Reviewing special considerations for CSPs

We will begin this chapter by understanding the licensing models for Azure Stack Hub.

Understanding Azure Stack Hub licensing models

There are two main components of the Azure Stack Hub software:

- **Cloud infrastructure**: Facilitates the system, including the portal
- **Services**: Customer workloads running on the system

You will only be billed for the services that are running on the Azure Stack Hub instance. Examples of billed services are virtual machines that have been deployed from the Azure Stack Hub marketplace and Azure services such as Azure App Service.

There are two ways in which the services can be licensed:

- Pay-as-you-use (consumption-based model)
- Capacity model

We will cover each of these different licensing models, starting with the pay-as-you-use consumption model.

Pay-as-you-use

The pay-as-you-use model has no upfront fees attributed to it, which means that you will only pay when you actively use a service. This model offers a continuous transaction experience that is identical to Azure's. The usage of every service is metered and automatically transmitted to Microsoft Azure commerce. This information is then integrated and billed alongside your Azure usage.

There is no initial upfront implementation fee for pay-as-you-use from Microsoft, although there is likely to be a charge from the **Original Equipment Manufacturer (OEM)** vendor who deploys the Azure Stack Hub instance.

You will not be billed for the standard virtual machines and software needed to power the actual Azure Stack Hub infrastructure. This means that all of the Cloud Infrastructure, Management, Security, and Identity services, as well as Networking and Service Fabric, are not invoiced to you. The following table describes how metering for the various services available on Azure Stack Hub is undertaken at the time of writing. All the services are standalone, which means that if you fire up Azure App Service, only the Azure App Service meter will be running:

Packaging	Service	Metering Units
Up-front licensing	Azure Stack Hub Initial Deployment	n/a – no upfront fees
Consumption-based fees	Cloud Infrastructure, Management, Security, and Identity	n/a – included
	Networking and Service Fabric	n/a – included
	Virtual Machines: Base VM	$/VCPU/month
	Virtual Machines with Windows Server	$/VCPU/month
	Blob Storage Service	$/GB/month (no transaction fee)
	Tables and Queues Service	$/GB/month (no transaction fee)
	Managed Disks	Disk/month
	Unmanaged Disks	$/GB/month
	Azure App Service	$/VCPU/month
	Event Hubs	$/core/hour

There are two options when running Windows Server virtual machines:

- **Native meters**: Utilizes native meters included in Azure Stack
- **Existing Windows Server licenses**: Utilizes Azure Stack Hub Base VM hourly meters

To implement SQL Server on Windows virtual machines, you can reuse existing SQL licenses, alongside the Windows virtual machines. You must make sure you have enough Windows Server core licenses to cover the entire Azure Stack Hub region, regardless of how many Windows Server virtual machines are deployed. For example, if you have 100 physical cores in your Azure Stack Hub infrastructure and you deploy 75 virtual cores in Windows Server virtual machines using your own licenses, you will still require 100 core licenses to cover the whole region. Azure Stack Hub pay-as-you-use services are available via **enterprise agreements** (**EAs**) or **CSPs** and are sold in the same way as Azure services. The other licensing model for Azure Stack Hub is the capacity model, which we will cover next.

Capacity model

The capacity model offers a more traditional licensing model to address the needs of customers who need to deploy Azure Stack Hub in disconnected scenarios, and cannot report usage to Microsoft. An annual subscription fee is required for all physical cores of your Azure Stack Hub. The capacity model is available in an **Infrastructure-as-a-Service** (**IaaS**) package ($144/core/year) or a **Platform-as-a-Service** (**PaaS**) package ($400/core/year). The IaaS package allows you to use the compute and storage services. The App Service package includes all the services in the IaaS package, plus Azure App Services (including Web, Mobile, Logic Apps, and Functions). The following table shows this package breakdown:

Packaging	IaaS Package	PaaS Package
Azure App Service		X
Azure Storage	X	X
Base Virtual Machine	X	X
Windows Virtual Machine	BYO License	BYO License
SQL Server Virtual Machine	BYO License	BYO License

When using the capacity model to run your Azure Stack Hub instance, you must ensure you have existing Windows Server or SQL Server licenses to be able to run Windows Server and SQL Server virtual machines.

The capacity model is only available with an EA and can only be ordered via the standard volume licensing channels. The capacity model does not have incorporated billing with Azure, and Azure monetary charges cannot be applied to the capacity model.

Now that we understand the different licensing models within Azure Stack Hub, we will learn how to purchase Azure Stack Hub.

Overviewing how to purchase Azure Stack Hub

We have covered the two licensing models that are available for Azure Stack Hub and in doing so, we started to touch on some of the methods of purchasing Azure Stack Hub. We have already mentioned the EA and CSP options, and we will explore these further in this section.

The Azure Stack Hub platform is purchased as an integrated system, as we covered in *Chapter 1, What Is Azure Stack Hub?*, when we talked about what Azure Stack Hub is. This integrated system includes the necessary hardware and software, as shown in the following diagram:

Figure 13.1 – Azure Stack Hub integrated systems

The physical hardware is purchased through an OEM vendor such as Lenovo, Dell, HPE, or Cisco. The OEM vendor will typically include professional services for deploying both the hardware and the associated Azure Stack Hub software, as well as a single point of contact for support. The actual usage billing of the Azure Stack Hub services is performed by Microsoft using one of the licensing models we looked at in the previous section.

The software component can be purchased by the end customer either directly using an EA or via a CSP. The CSP would purchase the license from Microsoft via the CSP program to allow them to resell these services, as shown in the following diagram:

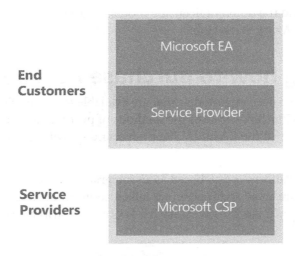

Figure 13.2 – Azure Stack Hub software purchase

The hybrid support experience offered in Azure Stack Hub covers the full system life cycle and is consistent and fully integrated.

You will need two support instruments to ensure that Azure Stack Hub is fully supported. The first is with Microsoft, who will cover the support for the cloud services. The second is with the hardware vendor who is offering the system support. The integrated support experience is designed to offer a consistent support experience, regardless of who you call first when it comes to coordinated escalation and system support.

Although support is sold as separate components, Microsoft and various hardware vendors have collaborated to establish a single integrated support experience. You only have to make one call to the vendor of your choice, be that Microsoft or the hardware vendor, for any Azure Stack Hub issue. That vendor will help you diagnose the source of the issue and route your queries accordingly to the other vendor, if and when necessary. As an example, if the OEM vendor is Lenovo, then the hardware that's supplied will be part of the ThinkAgile brand. For the ThinkAgile brand, Lenovo includes a support offering called ThinkAdvantage, which provides a single point of contact for any issue related to Azure Stack Hub. This is a dedicated Level 3 support team and if the issue is found to be software-related, then this team will raise the necessary call with Microsoft on behalf of the customer and continue to work with Microsoft until the issue is resolved. The customer would not have to raise their own call with Microsoft.

One of the key decisions when looking at purchasing Azure Stack Hub is around who is going to be managing the environment once it has been deployed. If the intention is to manage and maintain the environment yourself in your own data center, then you would typically buy the complete package from your hardware vendor, including both the hardware and software needed to run Azure Stack Hub. You would then manage and operate the platform going forward. If, however, you are not interested in managing the environment once it has been deployed, then you can purchase through a CSP. Typically, this means that the hardware is not deployed by the customer data center but by the service provider. The service provider is then responsible for the management and operation of the environment on your behalf. Some of the OEM vendors also double up as CSPs, while some will also offer a managed service for the platform, even when the Azure Stack Hub environment is run within the customer data center.

As well as purchasing Azure Stack Hub, the licensing of the software also needs to be taken into account, especially if the customer wishes to use their existing licenses.

Azure Stack Hub is treated the same as on-premises hardware for the purpose of Microsoft Azure Stack Hub licensing for existing software. Customers may apply licenses from any available purchasing channel (**EA**, **Services Provider License Agreement (SPLA)**, Open, and others) but they must comply with all the product licensing terms under whichever licensing agreement the software was obtained through. When any other software is used in collaboration with Azure Stack Hub, the fee structure is as follows: the licensing fees for the software (paid to the software vendor), plus any virtual machines required to establish the service.

Guidelines on how Microsoft Windows Server and SQL Server licensing are applied to Azure Stack Hub instances will be discussed in the following sections.

Windows Server licensing

It is possible to use existing Windows Server licenses instead of the native Windows Server hourly meters found in the pay-as-you-use model to deploy Windows Server virtual machines.

Windows Server licenses that are acquired outside of Azure Stack Hub are subject to terms and conditions stated in the Microsoft Product Terms for the licensing channel. The following are a couple of guidelines regarding how the licensing terms and conditions can be applied when existing Windows Server licenses are being used with Azure Stack Hub:

- **The number of licenses required for Windows Server being used with Azure Stack Hub**: To be compliant with Windows Server licensing, all the cores in an Azure Stack Hub region must be included, just as with licensing on Hyper-V. Additionally, all cores must be bound by the same edition of the license (all Datacenter or all Standard), since the virtual machine may be sitting anywhere on the Azure Stack Hub instance. Microsoft recommends Windows Server Datacenter for Azure Stack Hub since workloads are massively virtualized. You can use an EA, Open, or Select Plus Windows Server license here. Customers who use volume licensing licenses must also have enough **Client Access Licenses** (**CALs**) to cover the use case. Since Azure Stack Hub is on your own hardware and in your data center, you do not require **Azure Hybrid User Benefit** (**AHUB**) rights to use Windows Server in collaboration with Azure Stack Hub.

- **AHUB with Azure Stack Hub**: Azure Stack Hub is treated as an on-premises piece of hardware from a licensing viewpoint. As a consequence, you do not require AHUB to be able to use existing Windows Server licenses alongside committed Azure Stack Hub environments. Additionally, this AHUB benefit does not expand to bringing Windows Server EA licenses to a hosted, multi-tenant infrastructure – you cannot bring Windows Server EA licenses to these environments.

Note that Windows licenses can also be obtained from the OEM vendor who is supplying the hardware, as part of the configuration that is sold to the customer.

SQL Server licensing

Virtual machines that have been deployed to Azure Stack Hub that are running SQL Server can use SQL Server licenses that have been purchased outside of Azure Stack Hub, but they still remain subject to the Microsoft Product Terms. In addition to the SQL Server licenses, the Windows virtual machines themselves must also be covered by the Windows licenses.

The following are some guidelines that clarify how licensing terms and conditions are applied when using existing SQL Server licenses with Azure Stack Hub instances:

- **The number of core licenses required for SQL Server used with Azure Stack Hub**: It is possible to license SQL Server either based on the number of physical cores or by virtual machines. If you license based on the number of physical cores, you need to provide a license for the entire Azure Stack Hub region. If you chose to license based on virtual machines, you only need enough licenses to cover the number of virtual machines that are running SQL Server (subject to a minimum of four licenses per virtual machine). When licensing based on virtual machines, you may independently allocate SQL Server Enterprise and Standard edition licenses by using a virtual machine. As Azure Stack Hub runs on the customer's own hardware, you do not need License Mobility when using SQL Server under an EA on your own Azure Stack Hub hardware within your data center.

- **License Mobility**: Azure Stack Hub is considered an on-premises piece of hardware from a licensing viewpoint. As a result, you do not require License Mobility to use SQL Server licenses that have been committed to Azure Stack Hub environments. You will, however, require License Mobility if you bring your own SQL Server EA licenses to a service provider's multi-tenant hosted environment. In that situation, you must also establish that your service provider is an authorized License Mobility provider.

As with the Windows Server licensing, it may be possible to purchase the SQL Server licenses from the OEM vendor who is providing the hardware for your Azure Stack Hub environment.

Example scenarios

The examples shown in the following tables illustrate how services can be licensed in Azure Stack Hub. They demonstrate the contrasting licensing of using pure Azure Stack Hub meters against using on-premises licenses in conjunction with Azure Stack Hub.

The first table demonstrates the usage of native meters, where you only pay for what you use. This usage is metered on a minute-by-minute basis, and storage is decoupled from the virtual machine instances:

Raw Infrastructure	What is Used	What is Metered
100 Physical Cores	25 vCPU Windows Server VMs	25 vCPU Windows Server VMs ($/vCPU/min)
5 TB Attached Storage	50 vCPU Linux VMs	50 vCPU Linux VMs ($/vCPU/min)
	2 TB Azure Blob Storage	2 TB Azure Blob Storage ($/GB/mo)

If you decide to make use of your existing licenses when you deploy Windows Server virtual machines on Azure Stack Hub, you must bring your own license. As a result, you will only pay a consumption rate using Base VM meters. You must make sure that you have sufficient Windows Server core licenses to protect the whole Azure Stack Hub region, regardless of the number of Windows Server virtual machines you are going to deploy on the Azure Stack Hub region. In the scenario shown in the following table, 25 of the 75 virtual machine cores (vCores) are running Windows Server. However, as there are a total of 100 physical cores in the Azure Stack Hub region, 100 Windows Server core licenses are required. When used in collaboration with existing Windows Server licenses, Azure Stack Hub only uses consumption meters based on the Base VM rate of all the Windows Server virtual machines:

Raw Infrastructure	What is Used	What is Metered
100 Physical Cores	25 vCPU Windows Server VMs	Azure Stack Meters
5 TB Attached Storage	25 vCPU Linux VMs	50 vCPU Windows Server VMs ($/vCPU/min)
	25 vCPU SQL Server Standard	2 TB Azure Blob Storage ($/GB/mo)
	6 4-vCPU VMs	On-Premises Licenses
	1 1-vCPU VM	100 Cores EA Windows Server Datacenter
	2 TB Azure Blob Storage	TBD EA Windows Server CALs
		28 vCores EA SQL Server Standard

If you use existing SQL licenses to deploy the SQL Server virtual machines on Azure Stack Hub, you will pay for those SQL Server licenses, as well as the Windows virtual machines used to run SQL Server. As a result, as we have already included enough independently acquired Windows Server licenses to protect the entire Azure Stack Hub region, only a Base VM fee is used for the 25 vCores being consumed by the SQL Server virtual machines. If you are only intending to use SQL Server for a portion of your deployment, you may license it on a per-virtual machine basis. To comply with SQL Server licensing rules, there is a four-core licensing minimum per virtual machine. This means that when you deploy a one-node SQL Server virtual machine, you will still have to pay for and apportion four core licenses.

Service providers can offer Azure Stack Hub as part of their portfolio, which can enable Azure-consistent hybrid cloud services. The service provider must license the Azure Stack Hub instance from Microsoft through the CSP channel, and then provide you with finished services, value-added offerings, and support, just as they do for Azure.

This completes our overview of how to purchase Azure Stack Hub. Before we finish this chapter, we will cover some special considerations that need to be taken into account for CSPs.

Reviewing special considerations for CSPs

In this final section, we are going to review some special considerations that need to be thought about when considering Azure Stack Hub.

The Microsoft CSP program allows partners to completely own their end-to-end customer life cycle, which may include activities such as deploying new services, provisioning, management, pricing, and billing. The CSP program helps **value-added resellers** (**VARs**) and **managed service providers** (**MSPs**) to sell Microsoft software and cloud service licenses with additional support so that you can become more involved with your customer base. This means that every cloud solution, from Azure to Office 365, can be resold to your customers at a price you set, and with unique value that's been added by you. The goal of the CSP program is not simply to resell Microsoft services, but to enhance them and deliver them in a way that makes sense for you and your customers.

Service providers hold the end customer relationships, including billing and support. Azure Stack Hub tracks per-tenant usage using the same subscriptions as Azure and transmits data to Microsoft commerce. Azure Stack Hub's billing and usage are shown on the same invoices and partner tools as Azure. All CSP programs and rebates apply to Azure Stack Hub.

The following diagram shows the relationship between the CSP hosting and service providers:

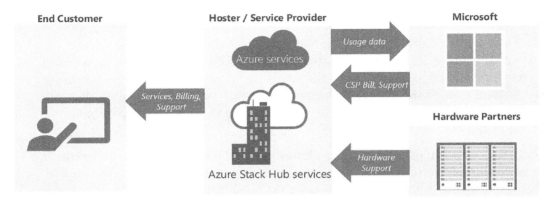

Figure 13.3 – Azure Stack Hub CSP

When looking at the Azure Stack Hub CSP model, you can see that there are two operating models you need to look at and understand. You need to think about and understand which CSP operating model works best for your particular organization. Let's walk through the two operating models to help you make this decision.

The first CSP operating model is the direct CSP model. In this particular operating model, the CSP operates Azure Stack Hub and has a direct billing relationship with Microsoft. All usage of Azure Stack Hub is directly billed to the CSP. In turn, the CSP will generate a bill for each customer or tenant who is consuming the services that are offered by the CSP. The period of billing, the amount to bill, and what you bill for are all controlled by the CSP. This model is shown in the following diagram:

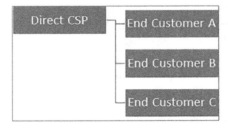

Figure 13.4 – Azure Stack Hub direct CSP operating model

The other operating model for CSP is the indirect model. In this model, the indirect CSP is also referred to as a distributor and is responsible for operating Azure Stack Hub. With this model, a network of resellers can help sell CSP services to the end customers. The indirect CSP has a direct billing relationship with Microsoft, and the usage of Azure Stack Hub resources is billed to the indirect CSP. The indirect CSP will then bill either the reseller or the end customer. This operating model is shown in the following diagram:

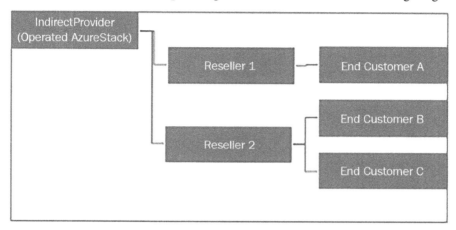

Figure 13.5 – Azure Stack Hub indirect CSP operating model

An indirect CSP has the same end customer responsibilities and subscription access rights that are granted via the Partner Center.

CSPs work with multiple customers using Azure Stack Hub. Multi-tenant billing debits each tenant's usage to its own Azure subscription. To achieve this, the following steps must be taken:

Figure 13.6 – Enabling multi-tenant billing for CSP

Two types of CSP subscriptions are available: **Azure Partner Shared Services (APSS)** and CSP subscriptions.

APSS subscriptions are the preferred method from Microsoft for registration when a direct CSP or a CSP distributor operates the Azure Stack Hub instance. This means that the CSP directly purchases the Azure subscription from Microsoft for their use. This creates the opportunity for partners to utilize a unified method for purchasing, tracking, and managing Azure, in addition to the ability to consolidate their Azure licensing and reseller agreements with Microsoft. APSS allows partners to have the same flexibility to utilize Azure subscriptions in CSP as they would with the Microsoft EA and Web Direct programs. This opens up scenarios such as building development and testing environments or deploying internal workloads, as well as hosting shared services or multi-tenant applications.

You can create an APSS subscription by going to the Microsoft Partner Center's administration portal. Once you've logged into the portal, navigate to **Settings | Account Settings | Shared Services**. If a **Shared Services** tenant does not already exist, then click on the **Create shared services** button, as shown in the following screenshot:

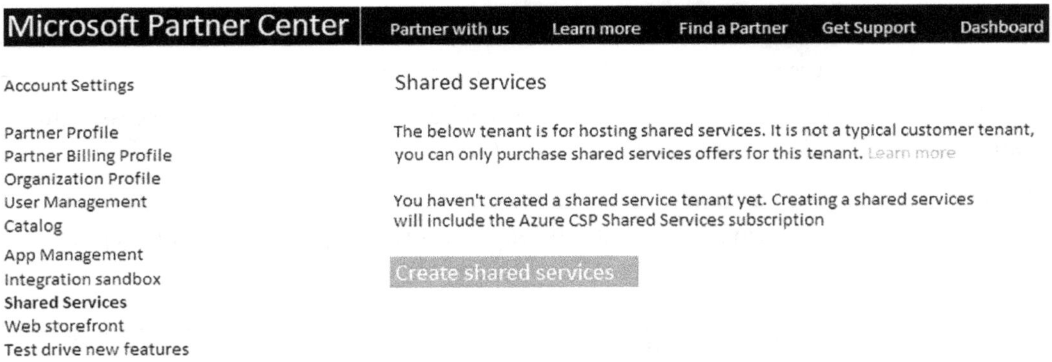

Figure 13.7 – Create shared services tenant in Microsoft Partner Center

This will create a shared services tenant that will purchase the Azure CSP Shared Services subscription. This will be utilized for shared resources and internal workloads.

The CSP subscription is the most common subscription model. Here, either a CSP reseller or the customer operates the Azure Stack Hub admin and tenant spaces, or in many cases, splits the responsibility, with the CSP managing the admin space and the customer managing the tenant space.

The following table details the various roles and responsibilities associated with the different models and subscription types:

Persona	Subscription	Azure Stack Operator	Usage and Billing	Selling	Support
Direct CSP	APSS	Direct CSP	Direct CSP	Direct CSP	Direct CSP
Indirect CSP or Distributor	APSS	Distributor	Distributor	Reseller	Distributor or Reseller
Reseller	CSP	Reseller	Indirect CSP or Distributor	Reseller	Distributor or Reseller
End Customer	CSP	End Customer	Indirect CSP or Distributor	Reseller or Distributor	Distributor or Reseller

Azure Stack Hub should be registered with Azure using an APSS or CSP subscription.

This completes our walkthrough of the Azure Stack Hub licensing models. We will build on these in the next chapter when we look at how usage and billing is handled for consumption. Before we move on, let's recap what we have learned in this chapter.

Summary

This chapter provided us with an overview of the licensing models within Azure Stack Hub. This introduced us to two different models; that is, the pay-as-you-use and capacity models. We looked at how Windows and SQL Server licensing is utilized in Azure Stack Hub, including the ability to use licenses bought outside of Azure Stack Hub. We then looked at the process of purchasing Azure Stack Hub, including via an OEM vendor. Finally, we looked at some considerations that need to be taken into account if Azure Stack Hub is managed by a CSP. This introduced us to both the direct and indirect CSP operating models. We then walked through the process of creating a partner account for registering for Azure Stack Hub and finished by looking at some roles and responsibilities across the various operating models.

This chapter has set us up for the next chapter, where we will look at consumption usage, tracking, and billing in greater detail. Please join me for the penultimate chapter.

14
Incorporating Billing Models

Following on from the previous chapter on licensing, we will now extend these licensing models to incorporate Azure Stack Hub's billing models. We will look at how consumption is billed through Microsoft Azure Stack Hub in both a consumption model and also a capacity model. As with the previous chapter, we will also look at some special considerations for the **cloud services provider** (**CSP**) model. We will also look at the tools that are available for getting usage reports out of Microsoft Azure Stack Hub.

This chapter will cover the following topics:

- Understanding billing models
- Reviewing billing for consumption
- Viewing and reporting usage data
- Understanding considerations for CSPs

We will begin this chapter by understanding the different billing models that are available in Azure Stack Hub.

Technical requirements

You can view this chapter's code in action here: `https://bit.ly/38eOGtv`

Understanding billing models

As with licensing, which we looked at in the previous chapter, there are two different models for billing. You can choose between the pay-as-you-use or the capacity billing model. Even if you select the pay-as-you-use model, these deployments must be able to report their usage through a connection to Azure at least once every 30 days. This means that the pay-as-you-use model can only be used with the connected deployment scenario. Let's look at each of these billing models in turn, starting with the pay-as-you-use model.

Pay-as-you-use

The pay-as-you-use billing model charges the usage to an Azure subscription. In this model, you only pay when you use the Azure Stack Hub services. To be able to use the pay-as-you-use billing model, you need an Azure subscription and the account ID associated with the Azure subscription. In most cases, an **enterprise agreement** (**EA**) subscription will be used by enterprise customers. CSP and CSP Shared Services subscriptions can also be used. Usage reporting is configured during the Azure Stack Hub registration process; we will cover this later in this chapter, in the *Viewing and reporting usage data* section. The CSP subscription that you should choose depends on the CSP scenario, as shown in the following table:

Scenario	Domain and subscription options
You are a direct CSP partner or an indirect CSP provider, and you will operate the Azure Stack Hub.	Use a CSP Shared Services subscription or create an Azure AD tenant with a descriptive name in Partner Center, with an Azure CSP subscription associated with it.
You are an indirect CSP reseller, and you will operate the Azure Stack Hub.	Ask your indirect CSP provider to create an Azure AD tenant for your organization with an Azure CSP subscription associated with it using Partner Center.

Now, let's look at the capacity model.

Capacity-based billing

The capacity-based billing model requires an EA Azure subscription for registration. This registration sets up the availability of items of the marketplace, which needs an Azure subscription. This subscription will not be used for Azure Stack Hub usage. A disconnected Azure Stack Hub can only use the capacity-based billing model as usage information cannot be shared with Azure.

The capacity-based billing model is dependent on an Azure Stack Hub Capacity Plan SKU, which must be purchased based on the capacity of the Azure Stack Hub system. To be able to order the correct quantity, you need to know the total number of physical cores in the Azure Stack Hub system.

With that, we understand the two billing model options available within Azure Stack Hub. This leads us nicely to the next section, where we will review how we bill for consumption in Azure Stack Hub.

Reviewing billing for consumption

In this section, we will look at how we bill for consumption. We will begin by looking at how we bill for consumption in Azure.

When using the pay as you use model in Azure, a user subscribes to an offer, which gets them a subscription ID. The user then uses Azure resources, and usage meters are submitted to Microsoft alongside the associated subscription ID. These details are then processed by Microsoft Commerce, who applies the prices to the quantities being consumed and generates the bill. Usage charges can be viewed using the Azure Account Center, as shown in the following screenshot:

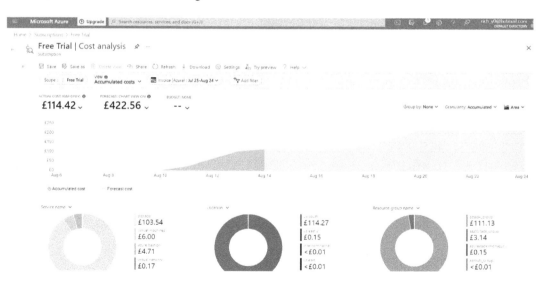

Figure 14.1 – Azure Account Center – Azure Stack usage

The billing process that occurs when a CSP is involved is slightly different from Azure. For this scenario, the partner will create a subscription for the tenant. The tenant then uses the Azure resources, and the usage meters are submitted alongside the associated tenant subscription ID. These meters are processed by Microsoft and prices are applied to the quantities. Microsoft then bills the CSP partner with details for each tenant. The CSP partner will then process this bill and split it into individual bills for each tenant.

Generally, the process of billing for consumption from Azure Stack Hub follows the same pattern we just described for Azure. In Azure Stack Hub, a user subscribes to an Azure Stack Hub offer, which generates them an Azure Stack Hub subscription ID. The user makes use of Azure Stack Hub resources, and usage meters are submitted to Azure Stack Hub with the associated subscription ID. Azure Stack Hub then forwards this usage meter data to Azure. This usage is then processed by Microsoft commerce, who generates the bill and sends it to the customer.

Azure Stack Hub is a private Azure cloud that is distinct from the public Azure cloud. This means that Azure Stack Hub has its own subscriptions that are different from the subscriptions in Azure. It also has its own usage meters, which, again, are different from the ones used in Azure. Azure Stack Hub is designed to work with Azure, and this includes metering and billing. The usage pipeline sends Azure Stack Hub usage to Azure commerce, and customers are charged via their Azure subscription.

Every resource provider in Azure Stack Hub reports usage data per resource usage. The usage service aggregates usage data periodically and stores it in the usage database. During Azure Stack Hub registration, you configure Azure Stack Hub to send this usage information through to Azure commerce. Once this data has been submitted, it will be visible in the billing portal. This is the usage pipeline; the following diagram displays its key components:

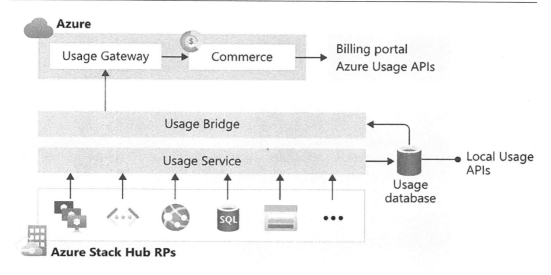

Figure 14.2 – Azure Stack Hub usage pipeline

We will cover the usage pipeline in the next section, where we will look at viewing and reporting usage data.

Viewing and reporting usage data

As we mentioned in the previous section, each resource provider available in Azure Stack Hub posts usage data per resource usage. The usage service periodically aggregates this usage data and stores it in the usage database. Azure Stack Hub operators and users can access this stored usage data through the Azure Stack Hub resource usage APIs.

If Azure Stack Hub has been registered with Azure, then, as stated previously, Azure Stack Hub will send this usage data to Azure Commerce. After this data has been uploaded to Azure, it is possible to access it through the billing portal or by using Azure resource usage APIs.

All Azure Stack Hub resource providers, including **Compute**, **Storage**, and **Network**, generate usage data at hourly intervals for each subscription. This usage data includes information about the resource being used, such as the resource's name, subscription used, and quantity used. After this usage data has been collected, it is reported to Azure to generate a bill, which can be viewed through the Azure billing portal.

The Azure billing portal shows all the usage data for chargeable resources. In addition to these chargeable resources, Azure Stack Hub also captures usage data for a broader set of resources, which you can access in the Azure Stack Hub environment through REST APIs or PowerShell cmdlets. Azure Stack Hub operators can get the usage data for all user subscriptions. Individual users will only be able to view usage data for their assigned subscription.

To set up usage data reporting, the Azure Stack Hub instance must be registered with Azure. When this registration process takes place, you can configure the Azure bridge component. The Azure bridge component is what connects Azure Stack Hub and Azure. The following is the usage data that is sent from Azure Stack Hub to Azure:

- **Meter ID**: Unique ID for the resource that was consumed.

- **Quantity**: Amount of resource usage.

- **Location**: Location where the current Azure Stack Hub resource has been deployed.

- **Resource URI**: Fully qualified URI of the resource that usage is being reported for.

- **Subscription ID**: Subscription ID of the Azure Stack Hub user, which is the local subscription.

- **Time**: Start and end time of the usage data. There is a delay between the time when these resources are consumed in Azure Stack Hub and when the usage data is reported to commerce. Azure Stack Hub aggregates usage data every 24 hours, and reporting usage data to the commerce pipeline in Azure takes another few hours. Therefore, the usage that happens shortly before midnight may not appear in Azure the following day.

It is possible to test usage data reporting by creating resources in Azure Stack Hub and leaving them to run for a few hours. The usage information is collected every hour and transmitted to Azure to be processed in the Azure commerce system. This process can take several hours to complete.

Where you can view this usage information will vary, depending on the type of subscription that has been used to register the Azure Stack Hub instance.

When Azure Stack Hub is registered against a CSP subscription, then the usage data for Azure Stack Hub can be viewed in the same way as normal Azure consumption. Azure Stack Hub usage is included in the invoice and in the reconciliation file, which is available in the Microsoft Partner Center (https://partnercenter.microsoft.com/partner/home). The reconciliation file is updated every month. Azure Stack Hub usage information can be accessed using the Partner Center APIs.

The reconciliation file can be downloaded from **Partner Center** and contains detailed usage data. The **AdditionalInfo** field contains the Azure Stack Hub subscription ID so that you can identify who consumed what resource.

The following screenshot shows Partner Center, including the reconciliation download link:

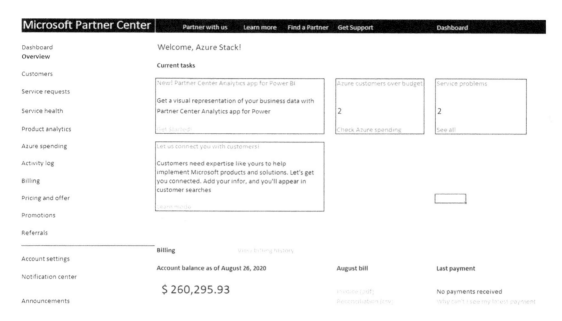

Figure 14.3 – Partner Center reconciliation file download

When Azure Stack Hub is registered using an EA subscription, then usage and charges are viewed within the EA portal (`https://ea.azure.com`). Azure Stack Hub usage is included in the advanced downloads, along with Azure usage, in the **Reports** section of the portal.

When Azure Stack Hub is registered using any other subscription type, then the usage and charges can be viewed in the Azure Account Center (`https://account.windowsazure.com/subsciptions`). The Azure account administrator can select the Azure subscription that was used to register the Azure Stack Hub instance, which will then display the usage data and amount that was charged for each of the resources.

Infrastructure virtual machines and virtual machines that are created as part of the deployment process are not included in the usage data that is transmitted and charged for.

If existing Windows licenses are being used, then meters are not generated for these machines.

Azure Stack Hub includes all the infrastructure required to track the usage as it happens and sends it to Microsoft Azure. Microsoft Azure commerce processes the usage data and adds usage charges to the appropriate Azure subscriptions. This is the same process that's used in the public Azure cloud for reporting and charging usage.

Most of the concepts for usage and billing are consistent between public Azure and Azure Stack Hub. Azure Stack Hub has a local subscription that is similar to an Azure subscription. Local Azure Stack Hub subscriptions are mapped to the corresponding Azure subscriptions when usage is sent up to Azure.

There are differences regarding how services are charged between Azure and Azure Stack Hub. As an example, the charge for virtual machines in Azure Stack Hub is based solely on the vcore/hour, and there is no concept of different size virtual machines as there is in Azure.

An Azure Stack Hub operator can utilize the provider usage API to view the usage of their direct tenants. This provider usage API can be accessed using both PowerShell or via the REST API.

To retrieve usage data via PowerShell, you can use the `Get-AzSubscriberUsage` PowerShell cmdlet, as shown in the following code example:

```
Get-AzSubscriberUsage -ReportedStartTime "2021-01-01T00:00:00Z"
 -ReportedEndTime "2021-01-02T00:00:00Z"
```

Alternatively, the REST API can be used by calling the `Microsoft.Commerce.Admin` service with a `Get` command, as shown here:

```
https://{armendpoint}/subscriptions/{subId}/providers/
Microsoft.Commerce.Admin/subscriberUsageAggregates?reportedStar
tTime={start-time}&reportedEndTime={end-endtime}&aggregationGr
anularity=Hourly&subscriberId={subscriber-id}&api-version=2015-
06-01-preview
```

These examples will return the usage for the given time period for the active tenant.

As a tenant with an active subscription, you can check on your usage by using the tenant APIs. As with usage in Azure, this can be retrieved using the `Get-UsageAggregates` PowerShell cmdlet.

The full list of Azure Stack Hub local meters can be found at `https://docs.microsoft.com/en-us/azure-stack/operator/azure-stack-usage-related-faq?view=azs2008`.

The following table represents a sample of some of the current local meters:

Resource Provider	Meter ID	Meter Name	Unit
Network	F271A8A388C44D93956A063E1D2FA80B	Static IP Address Usage	IP Addresses
	9E2739BA86744796B465F64674B822BA	Dynamic IP Address Usage	IP Addresses
Storage	B4438D5D-453B-4EE1-B42A-DC72E377F1E4	Table Capacity	GB*hours
	B5C15376-6C94-4FDD-B655-1A69D138ACA3	Page Blob Capacity	GB*hours
SQL RP	CBCFEF9A-B91F-4597-A4D3-01FE334BED82	Database Size Hour SQL Meter	MB*hours
MySQL RP	E6D8CFCD-7734-495E-B1CC-5AB0B9C24BD3	Database Size Hour MySQL Meter	MB*hours
Compute	FAB6EB84-500B-4A09-A8CA-7358F8BBAEA5	Base VM Size Hours	Virtual core hours
Key Vault	EBF13B9F-B3EA-46FE-BF54-396E93D48AB4	Key Vault Transactions	Request counts in 10k
App Service	190C935E-9ADA-48FF-9AB8-56EA1CF9ADAA	App Service	Virtual core hours

In addition to the standard metering and reporting that's available within the platform, there are also some third-party products available that can be used to manage consumption and billing in Azure Stack Hub. These include, but are not limited to, the following:

- http://www.cloudassert.com/Solutions/Azure-Stack: Cloud Assert Billing for Azure Stack Hub is a complete end-to-end metered billing tool for Azure Stack Hub that enables billing and chargeback within the Azure Stack Hub portal.

- `https://exivity.com/integrations/azure-stack`: Exivity provides a metering and billing solution for hybrid clouds, including Azure Stack Hub.

Now that we understand how to view and report usage in Azure Stack Hub, we need to look at some of the special considerations that need to be applied for CSPs to allow them to charge their customers in a multi-tenant scenario.

Understanding the considerations for CSPs

As a CSP, you will have multiple tenants who are using your Azure Stack Hub. Each of these tenants will have a CSP subscription in Azure, and you must direct the consumption from Azure Stack Hub to the relevant subscription in Azure.

To be able to charge your customers for the use of your Azure Stack Hub resources, then the following steps will need to be undertaken:

1. The CSP partner creates a CSP subscription for the tenant.
2. The CSP partner onboards the tenant to Azure Stack Hub.
3. The tenant obtains a local subscription.
4. The CSP partner adds the tenant subscription to the registration process.
5. The tenant uses Azure Stack Hub resources.
6. Azure Stack Hub forwards usage meters to Azure commerce.
7. Azure commerce bills CSP, with details per tenant.
8. CSP processes the details for tenant billing.

The first step is to create the CSP subscription, which is based on a shared services account type for Azure Stack Hub. There are two types of subscriptions that can be used while registering a multi-tenant Azure Stack Hub. These are as follows:

- **Cloud solution provider (CSP)**
- **Azure Partner Shared Services (APSS)**

If Azure Stack Hub is operated by either a direct CSP or a CSP distributor, then the preferred shared services account is the Partner Shared Services subscription. This APSS subscription is associated with a shared-services tenant. When registering for Azure Stack Hub, the credentials that are provided must be for an owner of the shared-services subscription.

This APSS account allows CSP partners to purchase Azure subscriptions for their own use, and also offers the same flexibility to use Azure subscriptions as they would with a Microsoft EA. This allows them to build development and test environments, deploy internal workloads, and, more importantly, host shared services or multi-tenant applications. An APSS can be created via the **Partner** portal and is available under **Settings** | **Account Settings** | **Shared services**, as shown in the following screenshot:

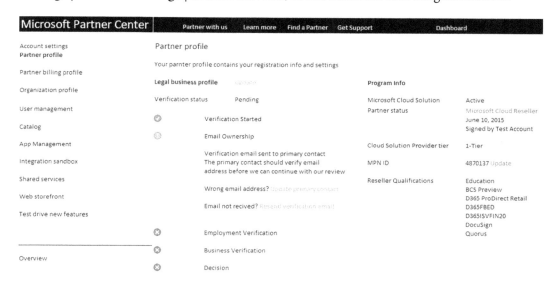

Figure 14.4 – Partner Admin – Shared services

If a shared services tenant does not already exist, then a new one can be created by clicking the **Create shared services** button, as shown in the following screenshot:

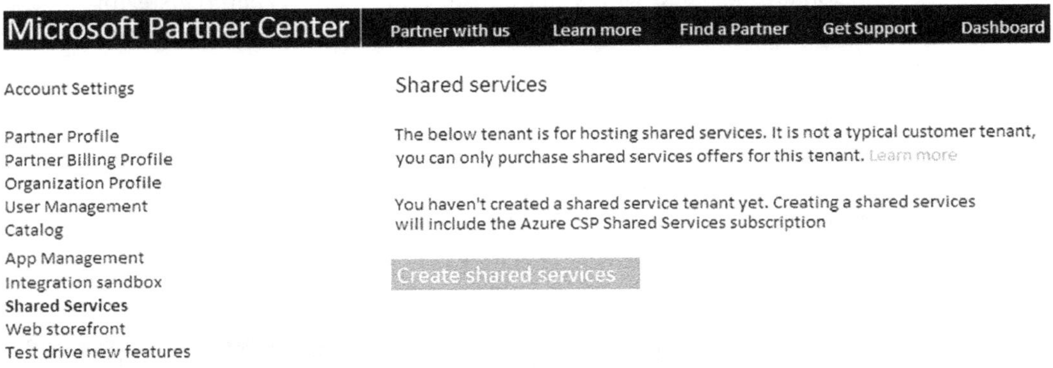

Figure 14.5 – Partner Admin – Create shared services

This will go ahead and create the shared services tenant and also purchase the Azure CSP shared services subscription, which can then be used for internal workloads and shared resources, as shown in the following screenshot:

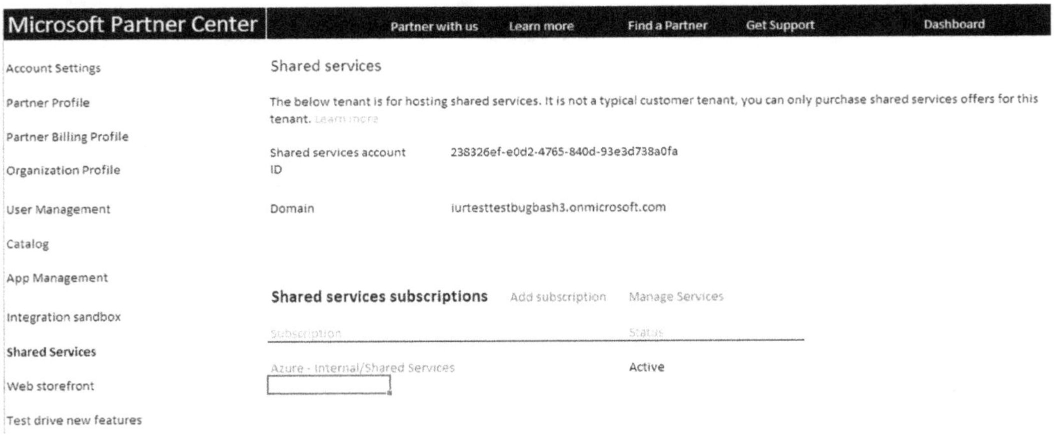

Figure 14.6 – Shared services subscription created

Once this subscription has been created, it can be used during the registration process for Azure Stack Hub, which will then set up the billing process.

If the Azure Stack Hub operator is a CSP reseller or an end customer, then the preferred subscription type is the CSP subscription.

We already covered the registration process as part of *Chapter 3, Azure Stack Hub Deployment*, when we walked through the deployment process, so we will not cover this again. However, once Azure Stack Hub has been registered against the correct subscription, we need to onboard the tenant subscriptions to ensure we can bill the correct subscription for the relevant usage.

Before you can add an end customer tenant to the Azure Stack Hub, you must enable multi-tenant billing on the Azure Stack Hub registration. This can be enabled by emailing `azstcsp@microsoft.com` with the registration subscription ID, resource group name, and registration name. It will take 1-2 days to enable multi-tenancy for this subscription. Once this is enabled, then the following steps need to be completed to onboard the tenant:

Figure 14.7 – Customer onboarding process

The first step, as shown in the preceding flowchart, is to create a new customer from within the **Partner Center** admin portal.

From within the **Partner Center** admin portal, select **Customers** from the menu and then click **Add customer**. This will open the **Account info** page, where you can enter details about the customer and their primary contact details. If you are an indirect provider, then you will also select the indirect reseller associated with this customer. Clicking **Next** will take you to the **Subscriptions** page, where you can select the subscriptions the customer is purchasing from you as the CSP. Clicking **Next** will take you to the **Review** page for you to double-check the information you have entered. Click **Submit** and then **Done** to create the new customer in the **Partner Center** admin portal.

The second step in the preceding flow is to create an Azure subscription for the end customer. If this was added during the initial creation from the **Subscriptions** page, then we can move on to the next step in the flow to create a guest user. If the subscription was not added to the customer during the creation process, then this can be performed in the **Partner Center** admin portal. Select **Customers** and then select the correct customer. Then, select **Add subscription** and select the appropriate subscription from the list for Azure Stack Hub. Add this subscription to the cart and then click **Review**. Once you are happy that you have the correct subscription, you can click **Buy** to purchase this subscription for the customer.

As a CSP, you would not typically have access to the customer's Azure Stack Hub subscription, but some customers would prefer the CSP to manage their resources. To enable this, the customer will need to add the CSP account as an owner to their Azure Stack Hub subscription. To be able to do this, they must add the CSP account as a guest user to their **Azure Active Directory (AAD)** tenant. It is recommended that an account that is not the Azure CSP's account is used to manage the customer Azure Stack Hub subscription.

Once you've done this, the Azure registration needs to be updated with the customer's subscription. Azure will then use the customer identity from **Partner Center** to report the usage. By doing so, we ensure that each customer usage is logged to the correct identity in Azure under the customer's CSP subscription. This ensures that the tracking and billing are split correctly across the multi-tenant platform. PowerShell can then be used to update the registration, as per the following command:

```
$ new-AzResource -ResourceId "subscriptions/
{registrationSubscriptionId}/resourceGroups/
{resourceGroup}/providers/Microsoft.AzureStack/
registrations/{registrationName}/customerSubscriptions/
{customerSubscriptionId}" -ApiVersion 2017-06-01
```

Note that this needs to be performed against all Azure Stack Hub instances that the customer uses. If they have access to multiple Azure Stack Hubs, then this step must be performed on each Azure Stack Hub.

The next step is to onboard the customer into Azure Stack Hub. To be able to do this, Azure Stack Hub must be configured to support multiple users from multiple AAD tenants by enabling multi-tenancy. These customer AAD directories will have a guest relationship with Azure Stack Hub. We covered the registration of these guest directories earlier in *Chapter 4, Exploring Azure Stack Hub Identity*, when we talked about identity.

Once the customer has been onboarded, then the last step in the workflow is to create a local resource in the customer tenant, such as a virtual machine, to confirm that security has been configured correctly and that billing can be tracked.

You can manage the tenant registrations in Azure Stack Hub by using PowerShell. For example, the following cmdlet will list all the registered tenants:

```
$ get-AzResource -ResourceId "subscriptions/
{registrationSubscriptionId}/resourceGroups/{resourceGroup}/
providers/Microsoft.AzureStack/registrations/
{registrationName}/customerSubscriptions" -ApiVersion 2017-06-
01
```

You can also remove a tenant from Azure Stack Hub using the following PowerShell cmdlet:

```
$ remove-AzResource -ResourceId "subscriptions/
{registrationSubscriptionId}/resourceGroups/
{resourceGroup}/providers/Microsoft.AzureStack/
registrations/{registrationName}/customerSubscriptions/
{customerSubscriptionId}" -ApiVersion 2017-06-01
```

With that, we've covered billing and consumption within Azure Stack Hub. Before we move on to the final chapter in this book, which is on support and troubleshooting, let's quickly recap what we have learned in this chapter.

Summary

In this chapter, we looked at the different billing models and how they related to our previous chapter on licensing. We talked about the two different billing models; that is, capacity and pay-as-you-use. We also looked at the usage pipeline and should now understand how Azure Stack Hub is metered. We now have an appreciation of the different meters that are available within Azure Stack Hub and how they are charged. We also know that usage data is transmitted to Azure commerce from Azure Stack Hub, and that this generates billing based on the resources that have been consumed. We know how we can view and report on this usage data from the platform based on the type of subscription that is in play for Azure Stack Hub. We also looked at what is required from a CSP to allow them to enable multi-tenancy and bill their tenants for the usage that they have consumed using the **Partner Center** portal.

We are now nearing the end of this book on Azure Stack Hub. In the next and final chapter, we will cover support and troubleshooting for Azure Stack Hub.

15
Troubleshooting and Support

This final chapter will cover the Microsoft Azure Stack Hub support model. We will walk through the integrated support experience and show a typical diagnostic flow for a support case. We will cover obtaining Microsoft Azure Stack Hub logs and engage with privileged endpoints. The final part of this book will then offer some tips on troubleshooting some common issues.

In this chapter, we will cover the following topics:

- Understanding the Azure Stack Hub support model
- Reviewing the integrated support experience
- Looking at Azure Stack Hub logs
- Understanding Azure Stack Hub troubleshooting

We will start this chapter by understanding the Azure Stack Hub support model.

Technical requirements

You can view this chapter's code in action here: `https://bit.ly/3Df88ob`

Understanding the Azure Stack Hub support model

Azure Stack Hub support is split into two parts – hardware support and software support:

- Hardware support is engaged directly with the hardware provider.

- Software support is engaged directly with Microsoft. If you already purchase software support from Microsoft (Azure, Premier, or Partner support), these contracts also cover Azure Stack Hub software support, and no additional contracts or fees are needed.

Although support is provided by the hardware vendor and Microsoft, the integrated support experience ensures a coordinated escalation and resolution so that you get a consistent support experience, no matter who you call first. We will cover this integrated support experience in the next section.

To understand the Azure Stack Hub support model, we need to look at where customers will buy their support and the contact breakdown as this will vary, depending on who owns and operates the Azure Stack Hub environment.

If the customer is consuming Azure Stack Hub via Microsoft using an **enterprise agreement (EA)**, then this enterprise customer will have two support contracts. The first one is with the hardware vendor, who will be responsible for the hardware and their own software that is used for Azure Stack Hub. The second support contract will be with Microsoft, which covers the Azure Stack Hub software and associated Azure services.

If the customer is a Cloud Solution Provider (CSP) then they only need to purchase one support contract from the hardware vendor. The support they receive from Microsoft will then depend on their CSP support level. The CSP will be responsible for level 1 and level 2 support for all Azure services, including Azure Stack Hub. The CSP is then backed up by Microsoft for level 3 support.

If the customer is consuming their Azure Stack Hub via a CSP, then, again, the customer will need to purchase two support contracts. The first contract will be with the hardware vendor, while the second will be with their CSP. As with enterprise customers, the hardware vendor will be responsible for the hardware and software used for Azure Stack Hub. The CSP is then responsible for level 1 and level 2 support for all Azure services, including Azure Stack Hub. As with the previous model, the CSP is backed up by Microsoft for level 3 support.

This covers the different support models for Azure Stack Hub. Next, we will review how support is handled with the integrated support experience.

Reviewing the integrated support experience

The hybrid support experience offered in Azure Stack Hub covers the full system life cycle and is consistent and fully integrated.

You need two support instruments to ensure that Azure Stack Hub is fully supported. The first is with Microsoft, who will cover the necessary support for the cloud services. The second is with the hardware vendor, who will offer system support. The integrated support experience is designed to offer a consistent support experience, no matter who you call first when it comes to coordinated escalation and system support.

As an example, when purchasing Azure Stack Hub from Lenovo, it comes with a support solution called **ThinkAgile Advantage**. This provides a single point of contact for all support issues related to Azure Stack Hub. The support engineers who are contacted through this process are level 3 engineers with a deep understanding of the Azure Stack Hub platform. This integrated support experience also means that if the issue is diagnosed as a Microsoft issue, then the Lenovo team will raise the issue with Microsoft on behalf of the customer. They will then work with Microsoft until the issue is resolved. This means that the customer will not have to initiate another call with Microsoft themselves, despite the support being purchased separately.

The following diagram demonstrates this integrated support experience:

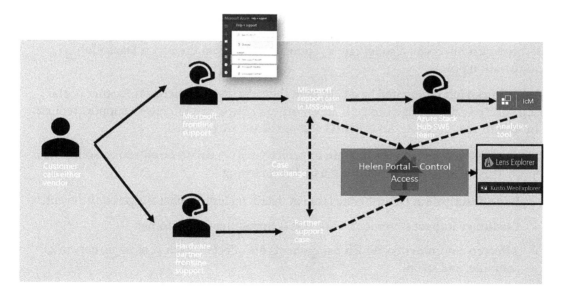

Figure 15.1 – Integrated support experience

This diagram demonstrates that the support experience for the customer should be consistent, regardless of whether the customer calls Microsoft for support on Azure Stack Hub or the hardware vendor who supplied the Azure Stack Hub environment.

With this experience, as mentioned earlier in this section, the hardware vendor is responsible for all hardware and any hardware vendor-branded software, including any issues with the **hardware life cycle host** (**HLH**). This includes any predictive monitoring and maintenance of items, such as disk drives and network cards. The hardware vendor is also responsible for deploying Azure Stack Hub, as well as any firmware updates and any hardware replacement issues. If the engineer from the hardware vendor runs into any issues during deployment, then they are responsible for level 1 and level 2 support but can call Microsoft for level 3 support. Any other issues are the responsibility of either Microsoft or the CSP. This can sometimes make it difficult to know who should be called first: the hardware vendor or Microsoft. The objective of the integrated support experience is that it should not matter who is called first, since the support should be given and the issue should be resolved regardless. Most hardware vendors, when providing support for Azure Stack Hub, will advise the customer to call their support first rather than Microsoft, regardless of whether the issue is hardware- or software-related. This is certainly true of Lenovo with their ThinkAgile Advantage support.

The way that a support request is raised will be dependent on how Azure Stack Hub is licensed. For example, enterprise customers can use the Azure portal (`http://portal.azure.com`) and click on **Help + support**. To be able to raise a support request via the Azure portal, a valid Azure subscription and support plan are required. Enterprise customers can also initiate a support request from the Azure Stack Hub administration portal.

If Azure Stack Hub is operated by a CSP, then they can use the Microsoft Partner portal (`https://partner.microsoft.com/en-us/support`) to raise a support request. To be able to use this route, an MPN ID and password are required.

When raising a support request, certain information is needed to be able to establish the support. This includes, but is not limited to, the following:

- **Issue description**: Describe the issue in detail, including what triggered the event.

- **Customer impact statement**: How this is impacting the customer.

- **Affected components**: Which components are affected, such as storage, network, compute, and so on.

- **Logs**: Provide any error messages, memory dumps, and screenshots.

The vendor who you raise the support request with is also likely to ask some questions along the following lines, to ensure that the support is directed correctly:

- Are you an Azure Stack Hub hardware partner?

- How many nodes are in the Azure Stack Hub system?

- What is the current patch level of the Azure Stack Hub system?

- What build number is currently running on the Azure Stack Hub system?

- What is the name of the Azure Stack Hub region?

- Is this Azure Stack Hub connected or disconnected?

- When did this issue start?

- Have you made any recent changes to the Azure Stack Hub system?

- Are you able to provide logs?

Some hardware vendors also provide proactive support through hardware monitoring, which can raise support requests without user intervention. For example, Lenovo includes a utility called XClarity within their servers and components. This controller monitors the health of the components, including disks and network cards, and also includes a call home function that will alert support of any issues with the hardware.

This covers the integrated support experience; we now have an understanding of how support requests can be raised. As part of this support process, as we mentioned previously, Azure Stack Hub log files may need to be sent to Microsoft or the hardware vendor to aid in troubleshooting. In the next section, we will look at these Azure Stack Hub logs and how we can share them.

Looking at Azure Stack Hub logs

As we mentioned in the previous section, there may be times when you're working through support issues with Microsoft or the hardware vendor, and they ask you to share diagnostic logs that have been created by Azure Stack Hub. These logs will have been generated by the Windows components within Azure Stack Hub and also the on-premises Azure services. The support personnel should be able to use the information from these logs to help resolve any issues with the Azure Stack Hub environment.

There are several ways that you can send this diagnostic information to Microsoft support, depending on the connectivity Azure has to Azure Stack Hub. Some of these options are as follows:

- **Send logs proactively**
- **Send logs now**
- **Save logs locally**

The flowchart shown in the following figure details which option is best for sending these diagnostics logs to Microsoft support based on the connectivity available. If Azure Stack Hub can connect to Azure, then it is recommended to enable proactive log collection. This will automatically upload the diagnostic logs to a Microsoft blob whenever a critical alert is raised on the Azure Stack Hub platform. In addition to this, you can send logs dynamically, on demand, by using the **Send logs now** option. If Azure Stack Hub is deployed in the disconnected scenario, then logs must be saved locally and then uploaded to Microsoft manually:

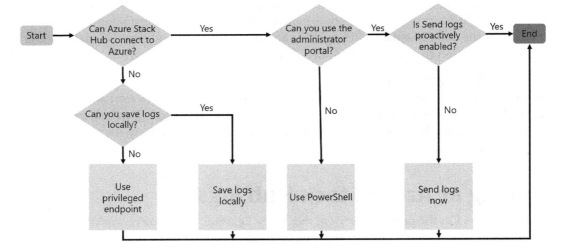

Figure 15.2 – Azure Stack Hub diagnostic logs collection

We will now walk through each of these options, starting with the **Send logs proactively** option.

Send logs proactively

The **Send logs proactively** option is the recommended option from Microsoft when the Azure Stack Hub platform is connected and registered with Azure. This option collects and sends the diagnostic logs from Azure Stack Hub to Microsoft automatically, even before a support case has been raised, based on critical events being generated.

With the most recent releases of Azure Stack Hub, the proactive log collection captures logs even if the event that triggered it is not visible to the Azure Stack Hub operator. This is done to ensure that the correct diagnostic information is collected and shared with Microsoft, without the need for an operator to intervene. This allows Microsoft support to begin their troubleshooting activities, even before the support case has been raised, which should lead to quicker problem resolution.

The proactive log can be enabled and disabled from the Azure Stack Hub portal. This can be found under the **Help + support** overview section. Then, select **Settings** to toggle this **On** or **Off**.

These automatically triggered periodic log collections are uploaded to a Microsoft-controlled storage account in Azure, and the data is based purely on the Azure Stack Hub system's health alerts. The data contained in these logs will only be used by Microsoft when troubleshooting system health alerts. The data that's contained in these logs may be stored by Microsoft in this Azure storage account for up to 90 days.

If **Send logs proactively** is not enabled, then it is possible to trigger this using **Send logs now**, which we will look at next.

Send logs now

The **Send logs now** option allows you to manually collect and upload your diagnostic logs from Azure Stack Hub to Microsoft when opening a support case. The easiest way to perform this operation is to use the administrator portal in Azure Stack Hub. This is the recommended approach from Microsoft when **Send logs proactively** is not enabled.

Assuming that the Azure Stack Hub platform is connected to Azure, the administrator portal is the best approach for sending the logs. However, if the portal is unavailable for any reason, then it is possible to perform this operation using PowerShell.

Sending the logs from within the administrator portal is straightforward and can be initiated from the **Help + support** section. Navigate to **Log collection** and then select **Send logs now**. You will be prompted for the start and end times of log collection. Select the relevant time window and then select **Collect and upload**.

If the portal is unavailable or you would prefer to use PowerShell, then the same operation can be performed using the `Send-AzureStackDiagnosticLog` cmdlet. The `FromDate` and `ToDate` parameters can be used to specify the time window, as per the portal command. If these parameters are omitted, then the log will collect diagnostic information for the last 4 hours only by default. There are several other parameters that can be used to filter the logs based on node, log type, resource provider, and role. A few examples are shown here:

```
Send-AzureStackDiagnosticLog -FilterByNode azs-node1
Send-AzureStackDiagnosticLog -FilterByLogType File
Send-AzureStackDiagnosticLog -FilterByResourceProvider eventHub
Send-AzureStackDiagnosticLog -FilterByRole VirtualMachines
```

As with the previous section, the data that's sent using this method is stored in an Azure storage account that is managed by Microsoft and, again, is retained for up to 90 days.

If Azure Stack Hub is deployed in the disconnected scenario, then the logs can be saved locally to allow them to be uploaded to Microsoft manually. We will look at this option next.

Save logs locally

It is possible to save the logs to a local **Server Message Block** (**SMB**) share when Azure Stack Hub is running in a disconnected scenario. This can also be done for a connected scenario if you are experiencing connectivity issues to Azure. Microsoft will provide detailed steps to transfer these saved logs when a support case has been raised. To be able to save logs to a share, an SMB path and user credentials must be entered in the **Settings** panel in the Azure Stack Hub portal, as shown in the following screenshot:

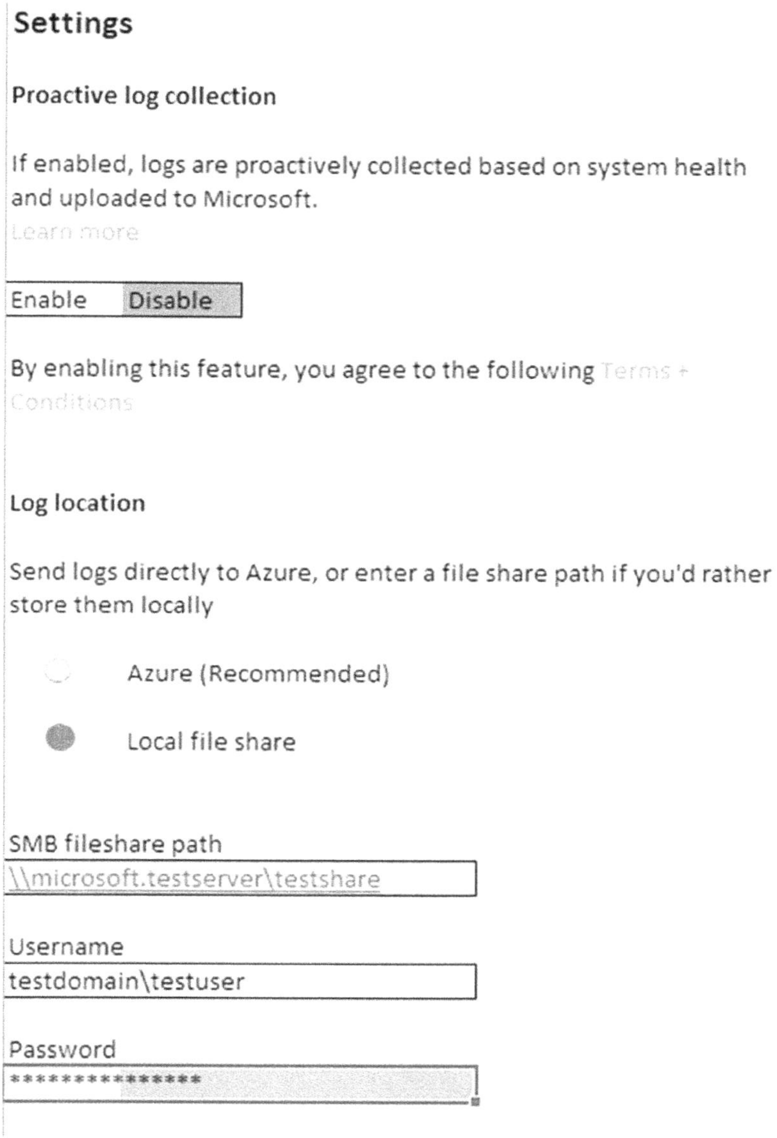

Settings

Proactive log collection

If enabled, logs are proactively collected based on system health and uploaded to Microsoft.
Learn more

| Enable | **Disable** |

By enabling this feature, you agree to the following Terms + Conditions

Log location

Send logs directly to Azure, or enter a file share path if you'd rather store them locally

○ Azure (Recommended)

● Local file share

SMB fileshare path
```
\\microsoft.testserver\testshare
```

Username
```
testdomain\testuser
```

Password
```
****************
```

Figure 15.3 – Azure Stack Hub portal settings – Local file share

If the administration portal is not available, then it is possible to save and send these logs from PowerShell using a **privileged endpoint** (**PEP**), which we will cover in the next section.

Sending Azure Stack Hub diagnostic logs by using a privileged endpoint

The `Get-AzureStackLog` PowerShell cmdlet must be run against a PEP on an integrated system. The following command is an example that can be run against the PEP to collect the necessary logs:

```
$ipAddress = "<IP ADDRESS OF THE PEP VM>" # You can also use
the machine name instead of IP here.

$password = ConvertTo-SecureString "<CLOUD ADMIN PASSWORD>"
-AsPlainText -Force
$cred = New-Object -TypeName System.Management.Automation.
PSCredential ("<DOMAIN NAME>\CloudAdmin", $password)

$shareCred = Get-Credential

$session = New-PSSession -ComputerName $ipAddress
-ConfigurationName PrivilegedEndpoint -Credential $cred
-SessionOption (New-PSSessionOption -Culture en-US -UICulture
en-US)

$fromDate = (Get-Date).AddHours(-8)
$toDate = (Get-Date).AddHours(-2) # Provide the time that
includes the period for your issue

Invoke-Command -Session $session { Get-AzureStackLog
-OutputSharePath "<EXTERNAL SHARE ADDRESS>"
-OutputShareCredential $using:shareCred -FilterByRole Storage
-FromDate $using:fromDate -ToDate $using:toDate}

if ($session) {
    Remove-PSSession -Session $session
}
```

This command will take some time to run based on which logs are being collected, the number of nodes, and the time window specified.

You can also use the `Invoke-AzureStackOnDemandLog` cmdlet to generate on-demand logs for particular roles. These logs are different from the logs that are generated by the `Get-AzureStackLog` cmdlet. Typically, you would only run this command under the guidance of Microsoft when asked to do so as part of troubleshooting a support case. The following command section provides an example of this `Invoke-AzureStackOnDemandLog` cmdlet:

```
$ipAddress = "<IP ADDRESS OF THE PEP VM>" # You can also use
the machine name instead of IP here.


$password = ConvertTo-SecureString "<CLOUD ADMIN PASSWORD>"
-AsPlainText -Force
$cred = New-Object -TypeName System.Management.Automation.
PSCredential ("<DOMAIN NAME>\CloudAdmin", $password)


$shareCred = Get-Credential


$session = New-PSSession -ComputerName $ipAddress
-ConfigurationName PrivilegedEndpoint -Credential $cred
-SessionOption (New-PSSessionOption -Culture en-US -UICulture
en-US)


$fromDate = (Get-Date).AddHours(-8)
$toDate = (Get-Date).AddHours(-2) # Provide the time that
includes the period for your issue


Invoke-Command -Session $session {
    Invoke-AzureStackOnDemandLog -Generate -FilterByRole
"<on-demand role name>" # Provide the supported on-demand role
name e.g. OEM, NC, SLB, Gateway
    Get-AzureStackLog -OutputSharePath "<external share
address>" -OutputShareCredential $using:shareCred -FilterByRole
Storage -FromDate $using:fromDate -ToDate $using:toDate
}


if ($session) {
    Remove-PSSession -Session $session
}
```

Diagnostic tools in Azure Stack Hub help with log collection and make it easy and efficient. Several components are used for diagnostic log collection, including something known as the **trace collector**.

The trace collector is run in the background to continuously collect **Event Tracing for Windows (ETW)** logs from the Azure Stack Hub components. These ETW logs have a 5-day age limit and can be located on a common local share. These logs will be deleted after 5 days, and new logs will be created. Each log has a default maximum size of 200 MB. The size of the current file is checked every 2 minutes and when the size limit is reached, it is saved and a new log file is generated. In addition to the limit of 200 MB per file, there is also an overall limit of 8 GB for the total file sizes generated.

The `Get-AzureStackLog` PowerShell cmdlet can be utilized to collect the logs from all the relevant components in Azure Stack Hub. The cmdlet saves the logs into a ZIP file that is located in a user-defined location. This is the `-OutputSharePath` parameter in the command example shown previously. Microsoft may request that the `Get-AzureStackLog` cmdlet is used when trying to troubleshoot a particular issue in Azure Stack Hub. It should be noted that these log files contain personal identifiable information, so they should be shared with caution.

It should also be noted that the size of the diagnostic logs collection can vary between the different methods that are used for collection. Proactive log collection, for example, will have an average size of 2 GB, while the average size of **Send logs now** will be dependent on the time period selected.

You can view the history of logs that have been collected from Azure Stack Hub over time on the **Log collection** page, which can be found in the **Help + support** section of the Azure Stack Hub administration portal. The information that's listed here includes the following:

- Time collected
- Status
- Logs start
- Logs end
- Type

An example of this information is shown in the following screenshot:

Collection Time ⓘ			Learn more		
Last 7 days		∨	Log collection overview ↗		
3 items					
COLLECTION TIME	TYPE	STATUS	FROM DATE	TO DATE	LOG UPLOAD SIZE
12/28/2019, 3:36:43 AM ...	Proactive	✓ Succeeded	12/28/2019, 1:53:21 AM ...	12/28/2019, 3:36:43 AM ...	3.51 GB
12/28/2019, 1:33:55 AM G...	On-demand	✓ Succeeded	12/27/2019, 5:33:47 PM ...	12/27/2019, 11:33:47 PM ...	5.42 GB

Figure 15.4 – Azure Stack Hub log collection history

All access to diagnostic data that's been shared with Microsoft is fully audited and logged. Microsoft employees will only be granted read-only access to these encrypted logs when they are involved in troubleshooting a particular support case. All data that's shared with Microsoft as part of this support diagnostic is deleted 90 days after the support case is closed.

This completes our overview of Azure Stack Hub logs and how they can be used to troubleshoot issues in Azure Stack Hub with Microsoft. We will now move on to the final section in this chapter, which is also the final section of this book. There, we will gain a basic understanding of troubleshooting in Azure Stack Hub.

Understanding Azure Stack Hub troubleshooting

Before raising a case with Microsoft support for issues with Azure Stack Hub, there are some commands that you can perform yourself to try to resolve some common issues. One of the first tools to utilize is the `Test-AzureStack` PowerShell cmdlet, which validates the Azure Stack Hub deployment and can help isolate common failures. You should always run `Test-AzureStack` before performing any system updates or applying hotfixes.

The `Test-AzureStack` cmdlet can be used to run infrastructure tests or cloud scenario tests against the Azure Stack Hub environment. To be able to perform the cloud scenario tests, you must use cloud administrator credentials. The cloud scenario tests are as follows:

- Creating resource groups
- Creating plans
- Creating offers
- Creating storage accounts

- Creating virtual machines
- Performing blob operations using the storage account that was created
- Performing queue operations using the storage account that was created
- Performing table operations using the storage account that was created

Microsoft may ask you to run this `Test-AzureStack` cmdlet from your management node and provide them with the logs that are generated. These logs should provide enough detail that you can focus on the area where an error is occurring.

One of the other common issues that you may come across is when resource providers are not registered. This will manifest itself when connecting to tenant subscriptions with PowerShell. You may notice that resource providers are not automatically registered. This can be resolved quickly through the use of the `Register-AzureRmResourceProvider` PowerShell cmdlet.

Some other common troubleshooting tasks that can be performed, without the need to involve Microsoft support, include resetting a Linux VM password when the reset password option is not working because of issues with the VMAccess extension. This can be reset as follows.

First, select a running Linux virtual machine to use as a recovery virtual machine. From the user portal, make a note of the virtual machine's size, **network interface card (NIC)**, public IP, **network security group (NSG)**, and data disks. Stop the virtual machine that is having an issue with the password. Remove this virtual machine and then attach the virtual machine as a data disk to the virtual machine you have chosen as your recovery virtual machine. Once this has been attached, sign into the recovery virtual machine and issue the following command:

```
sudo su -
mkdir /tempmount
fdisk -l
mount /dev/sdc2 /tempmount /*adjust /dev/sdc2 as necessary*/
chroot /tempmount/
passwd root /*substitute root with the user whose password you
want to reset*/
rm -f /.autorelabel /*Remove the .autorelabel file to prevent a
time consuming SELinux relabel of the disk*/
exit /*to exit the chroot environment*/
umount /tempmount
```

Back in the user portal, detach the disk from the recovery virtual machine. Then, recreate the original impacted virtual machine from the disk. You must ensure that you transfer all the information you made a note of earlier.

The other common issues we are going to cover in this section relate to the Azure Stack Hub patch and update process.

One of the common error messages you may encounter when attempting to install an update to Azure Stack Hub is "preparation failed." This normally indicates that the update failed to download, perhaps due to a problem with the internet connection or a firewall blocking the download. This can be attempted again by simply clicking **Install now**. If this does not resolve the issue, then the package can be manually uploaded and applied. The **Updates** blade is shown in the following screenshot:

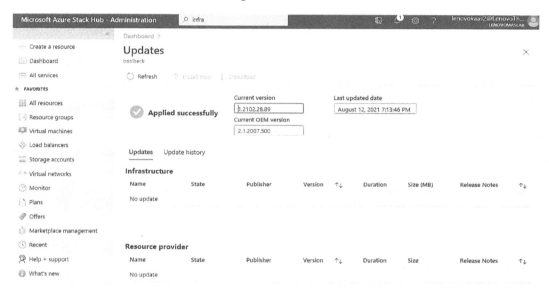

Figure 15.5 – Azure Stack Hub – Updates blade

Another error that is common when applying updates to Azure Stack Hub is the "update failed: check and enforce external key protectors on CSV" error. This is generally caused by the password for the **baseboard management controller** (**BMC**) not being correct. This can occur if the password has recently been changed, for example. Updating the BMC credentials and rerunning the update normally rectifies this error.

It is also very common to see error messages and alerts being reported in the portal while an update is being applied to Azure Stack Hub. This is normal behavior, and any messages that appear in the portal during the update about resources timing out and so on can safely be ignored until the update is complete. Azure Stack Hub includes the ability to retry tasks while it is processing in the event of intermittent failures, such as timeouts. Only if the status of the update is changed to failed does anything need to be investigated.

While we are talking about updates, it is worth noting that updates can be applied and monitored from the Azure Stack Hub administration portal. The **Updates** panel of the dashboard shows the current update version that is running. This information may well be needed when troubleshooting issues on Azure Stack Hub with Microsoft. The **Updates** panel allows you to view the high-level status of the update as it is processed and iterates through the Azure Stack Hub subsystems in turn. This also details additional information, such as steps completed and so on, while the Azure Stack Hub system is being updated. This blade will also provide the end status once the process has completed either successfully or failed. The **Update history** tab shows the updates that have been applied to this particular Azure Stack Hub environment, as shown in the following screenshot:

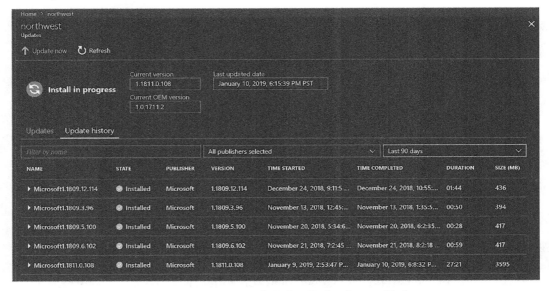

Figure 15.6 – Azure Stack Hub – Update history blade

It is important to apply Microsoft updates to Azure Stack Hub promptly to ensure that support from Microsoft is maintained. Microsoft recommend that Azure Stack Hub systems are running either the latest available release or one of the two preceding update versions. At the time of writing, the latest version that's available is 2008, and the two releases preceding this were 2005 and 2002. Any version before this, such as 1910, would be out of support, and Microsoft would ask you to update before they investigate any issues related to Azure Stack Hub.

This completes our walkthrough on the subject of support and troubleshooting for Azure Stack Hub. This also completes our journey with Azure Stack Hub in this book. Before we leave, let's recap what we have learned in this final chapter.

Summary

In this final chapter, we started by understanding the Azure Stack Hub support model. We learned that support is split between the hardware vendor who supplied the Azure Stack Hub platform and Microsoft. We now have an understanding of what support contracts are required based on how the Azure Stack Hub platform has been licensed and who provides that support. Then, we looked at the integrated support experience, which demonstrated that the support an Azure Stack Hub operator receives should be consistent, regardless of whether they call Microsoft or the hardware vendor for support. We looked at how incidents could be raised with Microsoft and how a typical support case would be handled. This then led us nicely into the next section, where we looked at the various ways of collecting diagnostic information from logs and sharing them with Microsoft to aid in troubleshooting. We should now be familiar with the different ways of sharing these logs, both through the portal and with PowerShell. Finally, we finished this chapter by looking at some common errors and troubleshooting tasks that can be performed by the Azure Stack Hub operator before contacting Microsoft for support. This included some common failures that can be encountered during the update process for Azure Stack Hub.

This rounded off this chapter nicely and also marks the end of this book on Azure Stack Hub. I hope this book has given you a great insight into Azure Stack Hub and has provided you with enough confidence to go on and tackle the AZ-600: Configuring and Operating a Hybrid Cloud with Microsoft Azure Stack Hub exam, which will gain you the Microsoft Certified: Azure Stack Hub Operator Associate certification.

Packt.com

Subscribe to our online digital library for full access to over 7,000 books and videos, as well as industry leading tools to help you plan your personal development and advance your career. For more information, please visit our website.

Why subscribe?

- Spend less time learning and more time coding with practical eBooks and Videos from over 4,000 industry professionals

- Improve your learning with Skill Plans built especially for you

- Get a free eBook or video every month

- Fully searchable for easy access to vital information

- Copy and paste, print, and bookmark content

Did you know that Packt offers eBook versions of every book published, with PDF and ePub files available? You can upgrade to the eBook version at packt.com and as a print book customer, you are entitled to a discount on the eBook copy. Get in touch with us at customercare@packtpub.com for more details.

At www.packt.com, you can also read a collection of free technical articles, sign up for a range of free newsletters, and receive exclusive discounts and offers on Packt books and eBooks.

Other Books You May Enjoy

If you enjoyed this book, you may be interested in these other books by Packt:

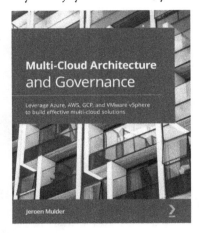

Multi-Cloud Architecture and Governance

Jeroen Mulder

ISBN: 9781800203198

- Get to grips with the core functions of multiple cloud platforms
- Deploy, automate, and secure different cloud solutions
- Design network strategy and get to grips with identity and access management for multi-cloud
- Design a landing zone spanning multiple cloud platforms
- Use automation, monitoring, and management tools for multi-cloud
- Understand multi-cloud management with the principles of BaseOps, FinOps, SecOps, and DevOps
- Define multi-cloud security policies and use cloud security tools
- Test, integrate, deploy, and release using multi-cloud CI/CD pipelines

Implementing Azure DevOps Solutions

Henry Been, Maik van der Gaag

ISBN: 9781789619690

- Get acquainted with Azure DevOps Services and DevOps practices
- Implement CI/CD processes
- Build and deploy a CI/CD pipeline with automated testing on Azure
- Integrate security and compliance in pipelines
- Understand and implement Azure Container Services
- Become well versed in closing the loop from production back to development

Packt is searching for authors like you

If you're interested in becoming an author for Packt, please visit `authors.packtpub.com` and apply today. We have worked with thousands of developers and tech professionals, just like you, to help them share their insight with the global tech community. You can make a general application, apply for a specific hot topic that we are recruiting an author for, or submit your own idea.

Share Your Thoughts

Now you've finished *Azure Stack Hub Demystified*, we'd love to hear your thoughts! Scan the QR code below to go straight to the Amazon review page for this book and share your feedback or leave a review on the site that you purchased it from.

`https://packt.link/r/1801078602`

Your review is important to us and the tech community and will help us make sure we're delivering excellent quality content.

Index

G

H

I

www.ingramcontent.com/pod-product-compliance
Lightning Source LLC
Chambersburg PA
CBHW081504050326
40690CB00015B/2916